WORLD RESOURCES INSTITUTE

THE COSTS OF CLIMATE PROTECTION:

A GUIDE FOR THE PERPLEXED

ROBERT REPETTO
DUNCAN AUSTIN

CLIMATE PROTECTION INITIATIVE

Carollyne Hutter
Acting Publications Director

Hyacinth Billings
Production Manager

Designed by:
Lomangino Studio Inc., Washington, DC

Each World Resources Institute Report represents a timely, scholarly treatment of a subject of public concern. WRI takes responsibility for choosing the study topics and guaranteeing its authors and researchers freedom of inquiry. It also solicits and responds to the guidance of advisory panels and expert reviewers. Unless otherwise stated, however, all the interpretation and findings set forth in WRI publications are those of the authors.

ISBN 1-56973-222-1
Library of Congress Catalog Card No. 97-60932
Printed in the United States of America on Recycled Paper.

CONTENTS

ACKNOWLEDGMENTS

The authors wish to thank many people who helped with the research underlying this report but not to suggest that they bear any responsibility for its contents. Richard Goettle, Richard Richels, Hugh Pitcher, Jae Edmonds, Robert Shackleton, Mike Shelby, and John Weyant were extremely helpful with data and technical suggestions. We thank John Reilly, Tom Walton, Alan Miller, and an anonymous external reviewer for useful criticisms and comments on an earlier draft of the report. Similarly, we thank WRI colleagues Paul Faeth, Jim Mackenzie, Walt Reid, Jonathan Lash, and Liz Cook for their comments. Valuable assistance at various stages from Marca Hagenstad, Theresa Bradley, Maggie Powell, Hyacinth Billings, Carollyne Hutter, Beth Behrendt, and Beth Harvey is greatly appreciated.

R.R.
D.A.

FOREWORD

Nations that have signed the Global Climate Convention are negotiating commitments to stabilize and then reduce emissions of greenhouse gases, which will otherwise continue to build up in the atmosphere and alter global climate. An agreement with binding limitations is essential, since experience in the United States and many other countries over the past five years shows that the purely voluntary efforts pledged in Rio de Janeiro at the Earth Summit are insufficient.

Since the Framework Convention on Climate Change was concluded in 1992, global emissions have continued to rise despite increasing evidence that human activity is having a discernible effect on world climate. The most recent scientific assessment by the Intergovernmental Panel on Climate Change emphasized that the continued buildup of greenhouse gases could have long-lasting climatic effects, some of which would impose significant economic burdens on nations and vulnerable populations.

Before the United States commits itself to specific restrictions on carbon dioxide emissions and timetables for implementation, it is essential that the economic consequences be thoroughly understood. Limiting carbon dioxide emissions will mean significant changes in energy use and energy sources, probably changing energy costs substantially. Household budgets and business profits will be affected. These impacts may affect inflation, international trade, patterns of investment, and thus the macro-economy. Whether the United States makes these changes unilaterally or in concert with other nations, whether it makes them in a cost-effective way using market-friendly policy instruments, and whether it implements them with adequate time and flexibility for economic adjustments to occur will all affect the macroeconomic impacts.

As the discussion of mandatory policies to reduce greenhouse gas emissions has accelerated, efforts to understand the economic implications of those policies have become intense. More than a dozen economic models have been used to generate simulations of different abatement targets and implementation policies. Not surprisingly, all this activity has produced little apparent consensus: economists derive markedly different predictions from their models about the likely impacts of achieving any specific abatement goal. Various interest groups have seized on particular predictions to support their own policy conclusions.

To sort out the resulting confusion, World Resources Institute vice president and senior economist Robert Repetto and his colleague Duncan Austin explain why different economic models reach different conclusions—because they start from different assumptions. This report, *The Costs of Climate Protection: A Guide for the Perplexed*, provides an overview of 16 leading economic simulation models. According to the report, under a reasonable set of common assumptions, models indicate that the macroeconomic impacts of stabilizing greenhouse gas emissions are likely to be modest and, if the environmental benefits are factored in, are likely to be beneficial. Repetto and Austin identify which assumptions are crucial. Their report will help readers form their own judgment about the likely impact of climate protection on the economy.

Building on the findings of earlier reports, including *The Right Climate for Carbon Taxes: Creating Economic Incentives to Protect the Atmosphere; Green Fees: How a Tax Shift Can Work for the Economy and the Environment;* and *Breathing Easier: Taking Action on Climate Change, Air Pollution and Energy Insecurity,* WRI continues to explore constructive ways to resolve the climate problem without undermining economic prosperity. We are committed to working with the private and public sectors in the United States and abroad to achieve this goal.

We would like to thank the MacArthur Foundation, the W. Alton Jones Foundation, the Nathan Cummings Foundation, The Energy Foundation, and the U.S. Environmental Protection Agency for providing generous financial support for the work underlying this report.

Jonathan Lash
President
World Resources Institute

INTRODUCTION

The latest international scientific assessment concludes that human activity is affecting the global climate (IPCC, 1996a). Population growth, rapid industrialization, increasing fossil fuel use, and continuing deforestation imply that without drastic action carbon dioxide concentrations in the atmosphere will double their pre-industrial levels by about the year 2060 and eventually cause average global temperature to rise by 1C-3.5C (IPCC, 1996a). The consequences of warming on this scale are hard to predict. Although some effects, such as rising sea levels and changing agricultural patterns, can be clearly foreseen, other effects, some potentially catastrophic, can only be guessed at. Nations have committed themselves to a precautionary approach designed to limit the accumulation of greenhouse gases.

Policies to prevent climate change focus mainly on carbon dioxide (CO_2) emissions, the most important greenhouse gas. As a first step toward the ultimate goal of stabilizing concentrations, the industrialized nations that signed the Framework Convention on Climate Change voluntarily undertook to return their emissions of carbon dioxide and other greenhouse gases to 1990 levels by the year 2000. However, many countries, including the United States, will fail to meet this target.

In 1995, at the first Conference of Parties to the Convention (COP-1) in Berlin, signatories acknowledged that even if emissions were stabilized at 1990 levels, concentrations would continue to rise rapidly because CO_2, once released, remains in the atmosphere for decades. The Berlin Mandate calls for strengthened commitments from developed countries to reduce their emissions after 2000. Negotiations since COP-1 have led to various proposals, the most stringent calling for industrialized countries to reduce emissions to 20 percent below 1990 levels by 2005 (AOSIS, 1995). The United States Government has stated that it is prepared to accept legally binding commitments in future protocols in the hope that this will prompt other countries to adopt a similar stance. Agreement on future commitments will be embodied in a formal protocol to be signed at the third Conference of the Parties in Kyoto in December 1997.

If the United States and other nations do agree on binding limits on greenhouse gas emissions, they will need to adopt measures to reduce carbon emissions with the least adverse economic impacts. One of the most effective and efficient mechanisms is a carbon tax—a tax levied on all fossil fuels in proportion to their carbon

One of the most effective and efficient mechanisms to reduce carbon emissions is a carbon tax—a tax levied on all fossil fuels in proportion to their carbon contents.

contents. Raising the prices of fuels and energy-intensive products would discourage all fossil fuel uses in proportion to their carbon contents and encourage development of less carbon-intensive alternatives. A recent statement by leading economists points out that a tax mechanism would be much more efficient than a regulatory approach (Economists' Statement on Climate Change, 1997).

An alternative proposal under serious consideration in the United States is a tradable permits program, in which permits would be required in order to sell or use fossil fuels. By limiting the total number of permits, the regulatory authority could control carbon emissions. If the government allowed permits to be bought and sold, the program would create efficient incentives like those of a carbon tax, because the permit price in the marketplace would signal how much firms should reasonably spend on abatement measures. If the government initially distributed the permits through an auction, it could mitigate adverse economic impacts by using the revenues to reduce other taxes without increasing fiscal deficits. In this sense, auctioned-off tradable permits to sell or use fossil fuels have economic implications similar to those of a carbon tax. (In this report, statements about the effects of a carbon tax apply equally to the impacts of tradable carbon permits that are auctioned off.)

Though one argument for tradable permits is the perceived political difficulty of proposing a change in the tax structure, the tradable permits approach also faces potential difficulties. It would be less efficient than a revenue-neutral tax and would encounter political opposition if valuable permits were given away to energy companies and utilities. Moreover, it would be difficult to include small fuel users in a tradable permits program, though their aggregate energy use is important, without creating administrative burdens much greater than those implied by raising energy taxes. Furthermore, if new scientific information necessitated further emissions reduction, canceling carbon permits that had been purchased in an auction or market transaction would be more difficult than raising a carbon tax.

Many interest groups claim that a carbon tax or any other efficient policy to reduce carbon emissions, such as a tradable permits policy, would impose high economic costs and reduce economic growth. For support, they point to simulations with economic models, some of which have suggested that stabilizing CO_2 emissions at 1990 levels could require a tax of up to $430 per ton of carbon by 2030 and could impose total costs of up to 2.5 percent of annual gross domestic product (GDP) (Charles River Associates, 1997). Of course, other economic models predict that similar emissions reductions could be achieved with far smaller energy taxes and negligible, or even favorable, overall impacts on the economy (Gaskins and Weyant, 1993).

Despite the complexity of the models, only a handful of easily understandable assumptions are important in determining the simulation results.

Which predictions should we believe—if any? Interested groups on different sides of the issues have their own preferred models (and modelers), and these tend to produce simulations supporting the policy positions of their respective sponsors. The underlying economic models may contain dozens of complicated equations that are nearly impenetrable to all but trained econometricians. How and why such models reach the predictions that they do is hard to comprehend. Yet, it matters greatly what the economic impacts of policies to reduce the long-term risks of global warming will be.

This report provides a guide for the perplexed—an explanation in simple terms of the key assumptions in the models being used to simulate the economic effects of carbon taxes or similar policies to control carbon dioxide emissions. The report also provides a quantitative analysis of 16 widely used models, demonstrating how key assumptions affect the predicted economic impacts of reaching CO_2 abatement targets. It turns out that despite the complexity of the models, only a handful of easily understandable assumptions are important in determining the simulation results. By showing the effect of these assumptions on the predicted economic costs, not just in one particular model but in all of them, this report can help readers to apply their own judgments about which models are more realistic and to reach their own conclusions about which economic predictions are more credible.

MODELS, ASSUMPTIONS, AND CONCLUSIONS

1

An economic model is no more than a coherent set of assumptions about the structure and functioning of the economy. A model is used to predict the consequences of some change, often a policy change like the imposition of a carbon tax. Naturally, the prediction depends entirely on the assumptions imbedded in the model—how could it be otherwise? Many of the assumptions of an economic model are simplifications, adopted to make the model easier to analyze or to compute. Modelers hope that in making these simplifying assumptions the baby is not disappearing along with the bathwater, but, alas, that is not always so. Many are based on empirical studies, often quite sophisticated, of particular relationships in the economy, and the modeler hopes not only that the relationship has been described accurately but also that it will continue in the future as it was in the past.

Many people are critical of the assumptions economists make but none more so than economists themselves. Typically, economic modeling of important issues is subjected to widespread and intense scrutiny within the economics profession, and unrealistic assumptions tend to be identified, improved, or discarded. Just as climate scientists and modelers over the past decade have criticized and improved the atmospheric models linking greenhouse gas emissions to changes in climate, so have economists improved the modeling of the economic impacts of a carbon tax. There has been prolonged economic debate and significant intellectual progress in making the models used for economic simulation more realistic. This report reflects some of that intellectual history.

Two kinds of assumptions in the models are critical: those that largely determine the predicted economic costs of abating carbon emissions and those relating to the economic benefits from forestalling environmental impacts from fossil fuel emissions. With respect to the costs of limiting carbon emissions, the key assumptions are

1. the extent to which substitution among energy sources, energy technologies, products, and production methods is possible;

2. the extent to which market and policy distortions create opportunities for low-cost (or no-cost) improvements in energy efficiency;

Just as climate scientists and modelers over the past decade have criticized and improved the atmospheric models linking greenhouse gas emissions to changes in climate, so have economists improved the modeling of the economic impacts of a carbon tax.

BOX 1 THE MAIN KINDS OF ECONOMIC SIMULATION MODELS

Predictions of the economic impacts of climate protection policies have been made on the basis of two main kinds of economic analyses, commonly referred to as 'top-down' and 'bottom-up' models. **Top-down** models are aggregate models of the whole economy that represent the sale of goods and services by producers to households and the reciprocal flow of labor and investment funds from households to industries. Models used for policy simulations also describe the role of government in imposing taxes, transferring income, and purchasing goods and services. *Computable general equilibrium* (CGE) models depict the formation of market-clearing prices in the process of matching the demand for goods and services from users to their supply by producers. Demand and supply conditions in such models are based on assumptions that consumers and producers allocate their resources to maximize their welfare or profits, respectively. However, demand and supply conditions are typically based on statistically estimated relationships observed in the past. *Optimizing* models derive their predictions by explicitly maximizing some assumed mathematical formula representing household welfare as a function of present and future consumption. Such general equilibrium models assume that households and industries will eventually respond efficiently to any policy change, though some models describe irreversible investment decisions and imperfect foresight regarding future prices that serve to delay the adjustment process.

In contrast, *macroeconomic* models predict economic behavior from statistically estimated relationships among economic variables in the past. Although such relationships are developed from accepted economic theory, macro models do not derive the predicted response to a policy shift from an explicit assumption that firms and households respond efficiently or with accurate foresight. Because they are estimated from actual macroeconomic behavior, they can reflect the short-term adjustment costs in response to an unexpected economic policy change, including business cycles, inflation, and unemployment. Macro and CGE models can be complementary in predicting short-run and long-run responses to a policy change. Moreover, modelers have learned to combine features of both (Hourcade and Robinson, 1996; Shackleton et al., 1992).

Top-down models used to analyze climate policies emphasize interactions between the energy sector and the rest of the economy. A tax-induced change in the price of carbon fuels directly affects demand and supply for energy, and indirectly affects other markets for commodities, labor, and capital. Therefore, consumer prices, incomes, savings, and labor supply are also affected, resulting in new levels of GDP, investment, and future growth. These can all be compared to baseline projections. More detailed analyses offer insights into distributional incidence on particular industries and income groups. When key assumptions are standardized among models, the range of their predictions narrows (Gaskins and Weyant, 1993).

Bottom-up analyses examine the technological options for energy savings and fuel-switching that are available in individual sectors of the economy, such as housing, transportation, and industry. Information on the costs of these options in individual sectors is then aggregated to calculate the overall cost of achieving a reduction in CO_2 emissions. In contrast to top-down models, in which the scope for technological substitution is extrapolated from past experience, bottom-up analyses estimate possibilities by considering explicitly the actual technologies that firms could profitably adopt at various energy price levels. Bottom-up analyses tend to be more optimistic about the scope for cost-effective energy savings.

To some extent, this optimism comes from overlooking important barriers to implementation, such as management and retraining time, risk-aversion toward unproven technologies, capital constraints, household preferences, or lack of information (Boero et al., 1991). Top-down models based on past rates of substitution and technological change implicitly incorporate such effects. Moreover, bottom-up analyses do not deal as adequately with overall macroeconomic constraints on capital availability or market demand as top-down models do. Despite these limitations, bottom-up analyses have highlighted energy inefficiencies and technological opportunities. Some top-down climate models have adopted features of bottom-up analyses by incorporating detailed descriptions of technological options for energy supply, conversion, and use.

Assumptions about the use of revenues generated from carbon taxes or from auctioning off carbon permits are crucial.

3. the likely rate of technological innovation and the responsiveness of such change to price signals;

4. the availability and likely future cost of non-fossil, backstop energy sources;

5. the potential for international 'joint implementation' of emissions reductions; and

6. the possibility that carbon tax revenues would be recycled through the reduction of economically burdensome tax rates.

Top-down models that assume limited substitution, slow technological change that does not respond to price signals, limited, expensive, or no availability of non-fossil energy sources, and the absence of international cooperation in achieving emissions reductions at least cost unfailingly predict that the economic costs of achieving any given carbon abatement target will be high. At the other extreme, bottom-up analyses that embody optimistic assumptions about the potential availability and rapidity of cost-effective, energy-saving technological improvements, and that neglect capital and other resource constraints can be counted on to predict low abatement costs.

In addition, assumptions about the use of revenues generated from carbon taxes or from auctioning off carbon permits are crucial. These revenues can be used to offset reductions in revenues if rates are cut on other taxes that are economically burdensome, without raising fiscal deficits. Many existing taxes on incomes, profits, and payrolls discourage savings, work, or investment by lowering after-tax returns to those activities. Economic studies suggest that lowering marginal tax rates for such existing forms of taxation and making up the revenue through a carbon tax would lessen the economic impacts of achieving a carbon abatement target. However, many early economic modeling simulations assumed that revenues from a carbon tax would not be used in this way but somehow returned in arbitrary "lump-sum" distributions to households, with no effect on incentives to work, save, or invest.

The final set of key assumptions concerns the environmental damages a carbon tax would avoid. Though averting these potential damages is the rationale for a carbon tax, most economic models are not constructed in ways that can take these damages into account. However, some models have factored in two types of savings:

1. avoiding the economic damages from climate change (the 'climate benefits'); and

2. reducing other air pollution damages associated with the burning of coal and other fossil fuels (the 'non-climate benefits').

The impact of climate change on the U.S. economy is a matter of great uncertainty, with predictions ranging from potential disasters—floods, hurricanes, droughts, and pestilence—to

Though averting potential environmental damages is the rationale for a carbon tax, most economic models are not constructed in ways that can take these damages into account.

potentially mild or even benign effects. Attempts at comprehensive assessment have projected that, on balance, climate change will impose net economic costs over the next century, rising with the extent and rapidity of the change in climate (IPCC, 1996b; Nordhaus, 1993; Cline, 1992). Some assessments have assumed a degree of risk-aversion that gives more emphasis to low-probability but severely damaging outcomes.

Few models have dealt with the potential environmental side-benefits of a carbon tax that would make coal and petroleum fuels more expensive and discourage their consumption. Since baseline projections predict that without a carbon tax coal burning in power plants and gasoline consumption in motor vehicles will increase in coming decades, air quality might deteriorate, harming human health and necessitating higher medical expenditures. A carbon tax would reduce these risks. Whether or not models take such environmental side-benefits into account substantially affects the economic impacts they predict.

As the next section will demonstrate, the divergent assumptions built into economic models in these key areas largely explain why their predictions regarding the economic costs of reducing emissions differ so widely. Under a reasonable standardized set of assumptions, most economic models would predict that the macroeconomic impacts of a carbon tax designed to stabilize carbon emissions would be small and potentially favorable.

Aside from the macroeconomic impacts, other considerations enter the debate over climate protection policies: notably, their distributional impacts and their effects on our international competitive position. As discussed in the final section of this report, the disproportionate impact of a carbon tax on low-income households now appears to be less than first thought and could be easily offset by other tax reductions or cost-of-living adjustments in social security and other transfer programs. By contrast, the disproportionate impacts on certain industries, particularly the coal mining industry and coal-carrying railway lines, would undoubtedly be substantial. To put this in context, the baseline projections against which these effects are evaluated predict sub-

Under a reasonable standardized set of assumptions, most economic models would predict that the macroeconomic impacts of a carbon tax designed to stabilize carbon emissions would be small and potentially favorable.

stantial expansion in coal mining in the western United States. The effects of a carbon tax designed to stabilize emissions at something like current levels would be largely to reduce the industry's rate of growth.

Concerns regarding the effects of carbon abatement policies on the international position of the U.S. economy have also weakened in force because of recent developments. The key issues, closely interrelated, are

1. whether reduced energy demand in the United States would help hold down world oil prices, improving our terms of trade;

2. whether higher domestic energy prices would stimulate energy-intensive industries in other countries to expand more rapidly; and

3. whether other countries would also adopt similar policies to restrict greenhouse gases, following the U.S. lead.

The United States is a sufficiently large importer of petroleum products that its demand affects world prices. Baseline projections imply that in the absence of a carbon tax petroleum consumption would continue to outgrow production capacity in non-OPEC regions, so that OPEC would supply an increasing share of the world oil market. By 2015, OPEC's share would exceed its peak two-thirds share in 1974, when it was able to raise energy prices sharply (U.S. EIA, 1996). A U.S. carbon tax that reduced U.S. energy demand could forestall increases in these prices and shift some of the economic impact on to foreign oil producers.

If, however, the United States alone imposed a significant carbon tax, international trade and investment in some energy-intensive industries might shift sufficiently to expand carbon emissions elsewhere and reduce U.S. production of those products. This now seems unlikely. Differential environmental policies appear to have a weak impact on trade and investment flows, if any (Repetto, 1995). Moreover, many non-OECD countries, including India, China, Mexico, and the republics of the former Soviet Union, have already raised energy prices unilaterally for purely economic reasons. Major OECD countries are likely to follow suit in instituting climate protection policies if the United States takes the lead. A few European countries have already enacted modest carbon taxes; others are seriously considering replacing some labor taxes with environmental taxes (OECD, 1997; Carraro and Siniscalco, 1996). The greater likelihood of coordinated international action means that adverse trade effects can be avoided.

HOW THE MODELS' ASSUMPTIONS DETERMINE WHAT ECONOMIC IMPACTS THEY PREDICT

2

To clarify how key modeling assumptions affect the predicted economic impacts of a carbon tax, we have assembled 162 different predictions from 16 of the most reputable and widely used economic models. Each of the models differs in its basic features and for each model, different simulation "runs" reflect either different policy assumptions (e.g., about the disposition of tax revenues) or different abatement targets and carbon tax rates. Each simulation "run" generates a predicted economic impact—measured here as the percentage change in GDP in some future year—and a corresponding percentage change in carbon dioxide emissions in the same year. Both variables are measured with reference to a baseline scenario, particular to each model, predicting what would happen in that year if no carbon tax or equivalent policy were adopted.

This measure of economic impact—future year GDP—is not ideal but was adopted because it is predicted in all models. If the predicted economic adjustment involves an initial slump from which the economy then recovers, a better measure would be the (discounted) loss of income and consumption over the entire period, but such a measure is not available in all models. More fundamentally, GDP is a measure of economic activity, not economic wellbeing: for one thing, it does not measure the nonmarket value of environmental quality.

The collection of model "runs"—plotted in Figure 1—shows how variable the predicted economic impacts are. For example, a carbon tax that induces a 35 percent reduction in CO_2 emissions could be expected to *raise* GDP over its projected baseline level by more than 1.5 percent or to *reduce* GDP by about 3 percent, depending on the economic models and modeling assumptions used. The majority of predictions suggest that abating CO_2 emissions will reduce economic activity and that eliminating a greater percentage of emissions will lower GDP more than proportionately. However, this apparent consensus does *not* imply that this prediction is likely to be accurate, but only that most modeling exercises have employed similar assumptions.

For each of these 162 modeling predictions, we have listed the main assumptions underlying the predictions. Some of these revolve around the basic features of the model; others refer to the policy options assumed for the specific simulation. Our list includes most of the key assumptions discussed in the previous section. Of the structural features of the models, the salient distinctions are:

1. Is the model of the CGE type, which assumes that the economy adjusts efficiently in the long-run, or is it a macro-model that assumes the economy suffers persistent transitional inefficiencies?

2. How much scope for inter-fuel and product substitution does the model assume, as indicated by the number of different energy sources and industrial sectors in the model?[1]

[1] This is just one indicator of potential substitution; the other, not so easily measured, is the ease with which one product can be replaced by another in response to changes in relative costs.

3. Does the model assume that one or more backstop non-fossil energy sources are available at some constant cost ?

4. How many years does the model assume to be available to achieve the specified CO_2 reduction target, expressed as a percentage reduction from projected baseline emissions in the final year?

5. Does the model assume that reducing CO_2 emissions would avoid some economic costs from climate change, or that no such costs exist?

6. Does the model assume that reducing fossil fuel combustion would avoid some damages from air pollution, or not?

FIG. 1

THE PREDICTED IMPACTS OF CARBON ABATEMENT ON THE U.S. ECONOMY (162 ESTIMATES FROM 16 MODELS)

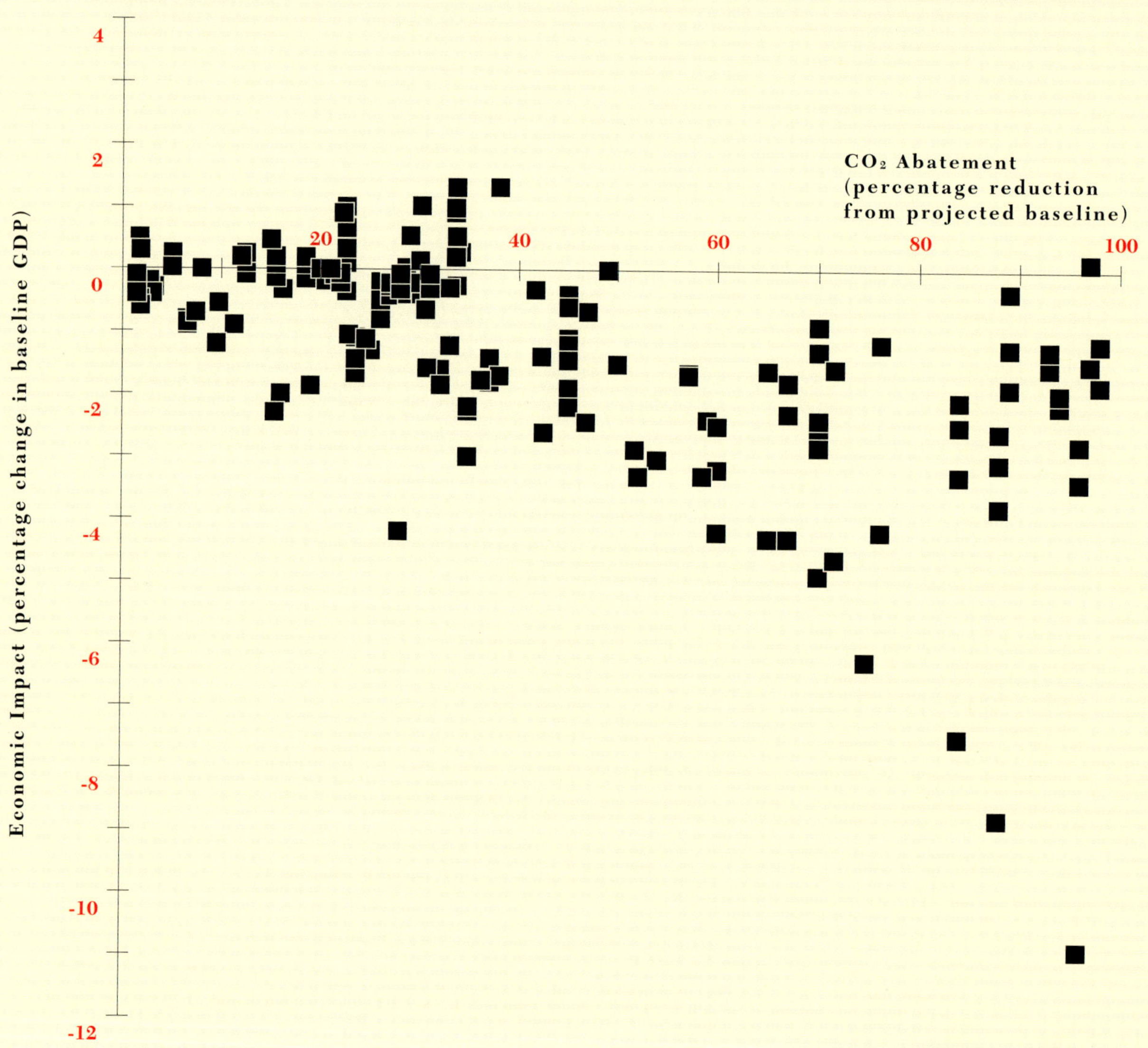

FIG. 2

PREDICTED IMPACTS ON GDP FOR VARIOUS CO_2 ABATEMENT TARGETS IN 2020: THE EFFECTS OF CHANGING UNDERLYING ASSUMPTIONS ONE-BY-ONE

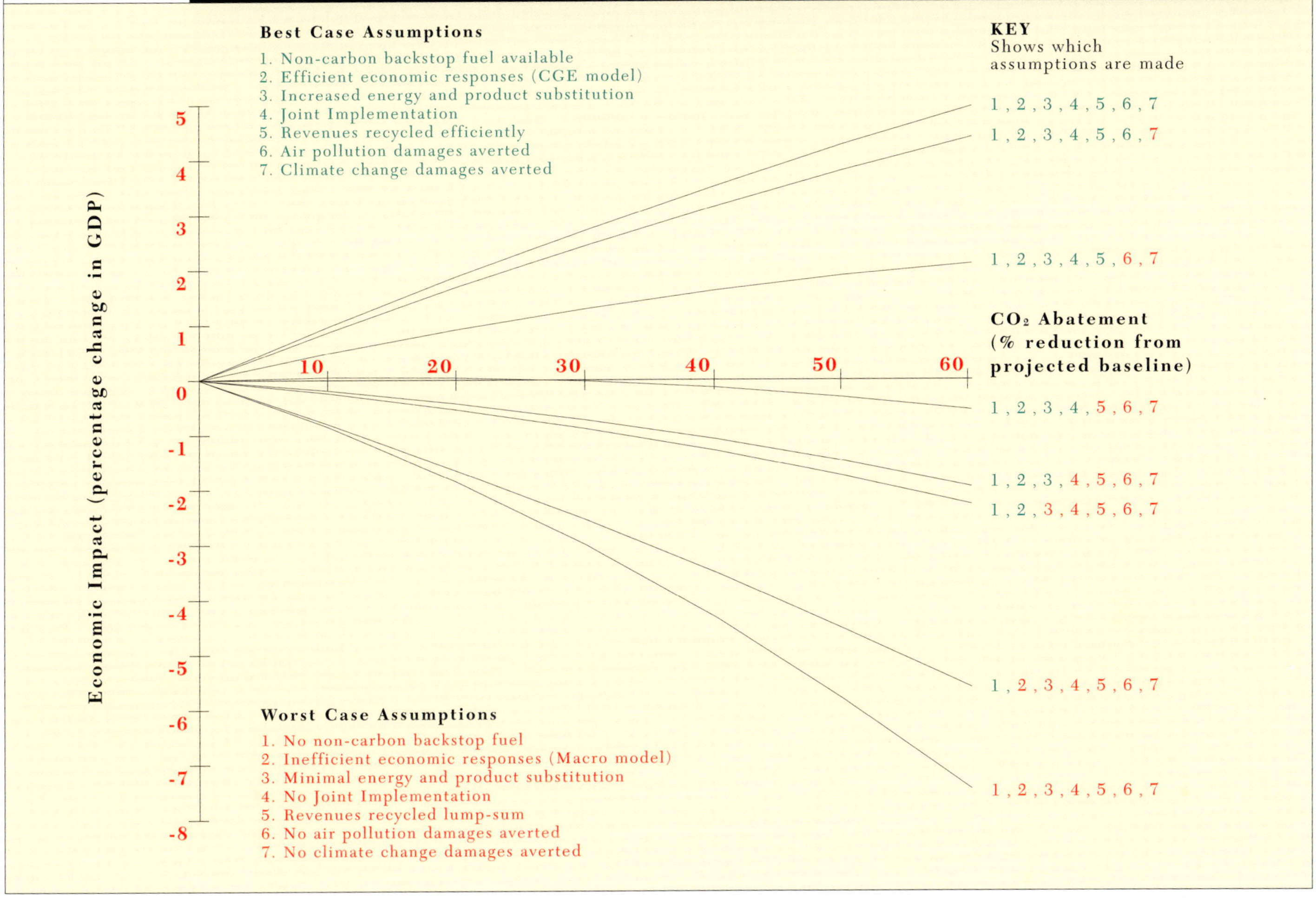

The salient policy assumptions that differentiate the model predictions are:

7. Does the model assume that carbon tax revenues are returned to the economy through the reduction of a distorting tax rate, or through lump-sum rebates?

8. Does the model assume that joint implementation options are available, or not?

To show how these assumptions affect the predicted economic impacts, we expressed these assumptions either as binary variables (yes = 1; no = 0) or numerical variables and used statistical techniques to relate the assumptions to the data points portrayed in Figure 1. Doing this shows how the assumptions affect the predicted impacts, not in any one model but across all 16 models in 162 different simulations. As suggested by these data points, we assumed that the economic impact of each additional one percent reduction in carbon emissions would be greater, the greater the percentage reduction.[2] We also assumed (consistent with the definition of the baseline projection) that imposing no carbon tax would not affect the economy's baseline trajectory, so that any statistical function would include the zero point in Figure 1.

Surprisingly, these eight assumptions (along with the size of the CO_2

[2] Specifically, we assumed that the percentage change in GDP was quadratically related to the percentage change in carbon emissions, both measured relative to the baseline projection.

emissions reduction) account for fully 80 percent of the variation in predicted economic impacts.[3] This is remarkable because it implies that all the other modeling assumptions—hundreds of assumed parameter values and relationships—are comparatively unimportant. Together, they account for only 20 percent of the differences among predicted impacts. Only a handful of basic assumptions really matters.

This is good news. People don't have to be Ph.D economists to understand the debate over the economic impacts of climate policy. Rather, people can use their own judgment and common sense to decide which of these basic assumptions are more realistic. Having decided that, they can then determine for themselves which predictions are more credible and what the economic impacts of a carbon tax or a climate stabilization policy are likely to be.

To illustrate, we have used the statistical relationship between predictions and assumptions to plot several cost curves in Figure 2. Each cost curve represents a different set of modeling assumptions selected from those listed above, starting from a set of "worst case" assumptions and then successively replacing them, one-by-one, with more favorable assumptions until a set of "best-case" assumptions is arrived at. In the statistical analysis underlying these curves, the slope of the curve connecting GDP change to emissions reduction was allowed to shift with each of the eight assumptions, but the year for achieving the abatement target was held constant.[4]

The worst case assumptions are that:

1. there is no non-carbon backstop energy source;
2. the economy does not respond efficiently to policy changes, even in the long-run;
3. the scope for inter-fuel and product substitution is minimal;
4. there is no possibility of joint implementation;
5. revenues are returned through lump-sum rebates;
6. there are no averted damages from air pollution; and
7. there are no averted damages from climate change.

Under these assumptions, many of which are obviously unrealistic, the adverse economic impacts of a carbon tax or equivalent policy would be severe, reaching 6 percent of end-year GDP for a 50 percent reduction in projected baseline emissions by 2020. *(See the bottom-most curve in Figure 2.)*

Scanning Figure 2 from the bottom up reveals the effects on predicted economic impact of changing these worst-case assumptions one-by-one. For

Surprisingly, these eight assumptions (along with the size of the CO_2 emissions reduction) account for fully 80 percent of the variation in predicted economic impacts.

[3] This was established by estimating a linear multiple regression model that incorporates variables representing the eight assumptions, taking the percentage change in GDP as the variable to be "explained". (The estimated regression equation can be found in the Annex.) An alternative regression analysis that included only the reduction in carbon emissions as an explanatory variable and excluded all the variables representing model assumptions explained only about half as much of the variation in predictions.

[4] The variable representing the presence or absence of a backstop energy source was specified to affect the curvature of the cost curve, since backstop sources become relevant only at higher energy prices and then limit the rate of cost increase.

Under all the best-case assumptions, a reduction in CO_2 emissions by 2020 would result in a substantial improvement in GDP relative to its business-as-usual path.

example, assuming that backstop energy sources exist improves the predicted economic impact substantially—by about one percent of GDP for a 50 percent emissions reduction. In Figure 2, what is notable about the predicted impacts on the U.S. economy is that changing only five worst-case assumptions—by assuming backstop energy sources, efficient long-run adjustment in the economy, greater substitution possibilities, joint implementation, and recycling of carbon tax revenues by reducing other burdensome tax rates—dramatically alters the predicted economic impacts. Instead of a six percent loss of GDP by 2020, there would be modest positive impact on GDP relative to the business-as-usual scenario.

Judging from all these simulations using a wide variety of economic models, the doomsday prediction of heavy economic losses if carbon emissions are reduced is implausible. It is more reasonable to predict that with sensible economic policies and international cooperation, carbon dioxide emissions can be reduced with minimal impacts on the economy.

Going further, Figure 2 indicates that if reducing fossil fuel combustion avoids economic damages from climate change or air pollution, then the overall economic impacts could be favorable.[5] The top-most curve in Figure 2 indicates that under all the best-case assumptions, a reduction in CO_2 emissions by 2020 would result in a substantial improvement in GDP relative to its business-as-usual path.[6] Of course, there is some degree of emissions reduction beyond which the incremental abatement costs exceed the value of the environmental damages that more pollution would create. This turning point is not adequately reflected in Figure 2, which should not be interpreted to suggest that if some carbon abatement is good, more is necessarily better. Figure 2 does imply, however, that models that take the environmental benefits of carbon taxes into account predict substantially more favorable economic impacts than models that ignore such benefits.

One target that has been analyzed extensively by the Interagency Analytical Team in preparation for the COP-3 meeting in Kyoto in December 1997 is a freeze on carbon emissions at 1990 levels by 2010 and stabilization of emissions thereafter. For the United States, it has been estimated that this target implies about a 26 percent reduction below projected baseline emissions in 2020, if the baseline is calculated on the basis of policies now in place (U.S. EIA, 1996). Figure 3 uses the same statistical analysis to show in detail the range of predicted long-run economic impacts if this target is attained.[7] Under unfavorable assumptions, GDP would be 2.4 percent lower in 2020 than under baseline conditions; under favorable assumptions, GDP would be 2.4 percent higher. Figure 3 also quantifies the relative importance of several modeling assumptions in creating this range of predictions.

Four assumptions stand out in terms of magnitude:

[5] This prediction is consistent with the interpretation of a carbon tax as a corrective tax that reduces a market failure—namely, the unintended effect of carbon emissions on the global climate. Economists agree that, if set at the proper rate, a tax to correct a market failure should improve an economy's productivity.

[6] When air pollution damages are assumed, an expanded measure of GDP in which environmental damages are recorded is the relevant indicator of economic impact.

[7] This analysis cannot encompass short-run transitional impacts predicted by some macroeconomic forecasting models.

FIG. 3

THE PREDICTED IMPACTS ON GDP IN 2020 OF STABILIZING CO_2 EMISSIONS AT 1990 LEVELS: THE EFFECT OF CHANGING UNDERLYING ASSUMPTIONS

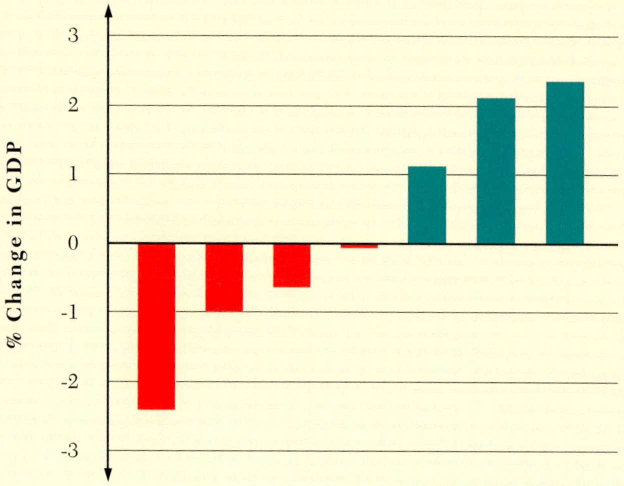

Worst Case Assumptions	↑	↑	↑	↑	↑	↑	↑	Best Case Assumptions
1. Inefficient economic responses*	1	1	1	1	1	1	1	1. Efficient economic responses*
2. No non-carbon backstop fuel	2	2	2	2	2	2	2	2. Non-carbon backstop fuel available
3. No Joint Implementation	3	3	3	3	3	3	3	3. Joint Implementation
4. Revenues recycled lump-sum	4	4	4	4	4	4	4	4. Revenues recycled effectively
5. No air pollution damages averted	5	5	5	5	5	5	5	5. Air pollution damages averted
6. No climate change damages averted	6	6	6	6	6	6	6	6. Climate change damages averted

* Macro Model

* CGE Model

- whether the economy will adapt efficiently;
- whether international joint implementation will be achieved;
- whether carbon tax or permit auction revenues will be recycled by reducing other taxes; and
- whether there will be economic benefits from abating pollution.

Most economists believe that the U.S. market economy, with high mobility of capital and labor, can adapt efficiently to moderate the impacts of policy changes. There is general agreement that a carbon tax that discouraged coal use in electricity generation would have the effect of reducing air pollution, even with current air pollution regulations in place. Whether to use revenue-raising policy instruments to limit emissions and how to dispose of resulting revenues are decisions that the U.S. government must make. Finally, international cooperation in joint implementation of carbon reduction targets is a possibility subject to negotiation. Under reasonable assumptions, the predicted economic impact of stabilizing emissions at 1990 levels would be neutral or even favorable.

MORE DETAIL ON THE KEY ASSUMPTIONS

3

The preceding analysis looked broadly at the key assumptions that turn out to determine very largely the predicted economic impacts of climate protection policies. These broad distinctions among the modeling assumptions explain most of the differences among predictions. Nonetheless, other aspects of the key assumptions, which could not be adequately built into the preceding analysis need to be recognized and understood. This section addresses such issues.

A. THE SCOPE FOR REDUCING ENERGY INEFFICIENCIES

Top-down models typically assume that all cost-effective improvements in energy efficiency have already been realized, an assumption contradicted by actual experience (DeCanio, 1993). For example, large companies that joined the Environmental Protection Agency's voluntary Green Lights Program to reduce their energy use found numerous opportunities to save both energy and money in their ongoing operations.

Bottom-up studies have found inefficiencies in energy use that could be remedied through building improvement measures, such as better insulation and low-energy lighting; through technological advances in transport efficiency; or through conversion of industrial processes. Assessments based on engineering studies suggest that from 20 to 25 percent of existing carbon emissions could be eliminated at an overall cost savings and that substantial further cutbacks could be made at relatively low cost (IPCC, 1996b; National Academy of Sciences, 1991; Office of Technology Assessment, 1991).

Some of these inefficiencies undoubtedly persist because of energy market imperfections, such as the divergence in incentives between tenants and landlords, builders and home purchasers; because of energy subsidies; or because of suboptimal decision-making within organizations. However, some reported savings opportunities might be illusory if the management costs of locating and implementing energy investments were overlooked or the differences in product and service characteristics of various energy-conversion technologies were ignored. Energy service companies, which seek to find and implement energy-saving opportunities on a contractual basis, have not found unlimited business opportunities at current low energy prices.

Top-down models typically assume that all cost-effective improvements in energy efficiency have already been realized, an assumption contradicted by actual experience.

Top-down models typically assume away energy subsidies that may encourage excessive fuel use and must be financed through higher levels of economically burdensome taxes. Energy subsidies, though not as prevalent in the United States as in some other countries, still include favorable tax and credit treatment for energy producers, below-market provision of power from public sector installations, and federally sponsored research and development. Two recent studies quantify annual federal energy subsidies at between $4.9-$14.1bn and $21-$36bn respectively (Alliance to Save Energy, 1993; U.S. EIA, 1992). Some of these subsidies, such as federally subsidized hydropower, actually reduce carbon emissions by replacing fossil fuels with hydroelectricity. Others, such as tax breaks for independent oil drillers, have no effect on U.S. oil consumption or carbon emissions, but merely replace foreign produced oil with domestically produced oil. Nonetheless, a study on the effects of removing U.S. energy subsidies concluded that significant reductions in CO_2 emissions could be achieved at no cost to the economy (Shelby et al., 1995).

B. SUBSTITUTION EFFECTS

A carbon tax will raise the price of fuels in proportion to their carbon content, increasing coal prices more than oil or gas prices and having little direct effect on hydro or nuclear power costs. Higher fuel prices will induce firms and households to seek ways to mitigate the cost increases. In particular, they will tend to

- substitute less carbon-intensive fossil fuels, such as gas, for carbon-intensive fossil fuels, such as coal (intra-fossil fuel substitution);
- substitute non-fossil energy sources for fossil fuels (non-fossil fuel substitution);
- substitute other factors of production (materials, labor and capital) for energy; and
- substitute less energy-intensive goods for energy-intensive goods (Cline, 1992).

The easier these substitutions are, the lower the overall burden of reducing CO_2 emissions. In addition, all such substitutions become easier as the time for adjustment increases. For example, in the short-term, a firm's production technique will be constrained by its existing equipment, but as new equipment and processes are brought on line, energy use can be reduced more readily. Similarly, consumers need time to replace durable goods and fully adapt purchasing habits to altered prices.

Economic models differ in the degree to which they represent these substitution possibilities. Highly aggregated models, which might have only a single producing sector (i.e., a sector producing a composite commodity called GDP), cannot incorporate the possibility of substituting one product for another. Similarly, models that recognize only two primary fuel sources cannot adequately represent inter-fuel substitution possibilities. More disaggregated models, such as the Markal-Macro model, which recognizes 11 primary fuel sources and dozens of fuel conversion technologies, are potentially better able to deal with such substitution possibilities (U.S. DOE, 1996).

However, this potential may or may not be realized. Despite being more or less disaggregated, models differ in the assumed ease of substitution among products and technologies in response to cost changes. Some models assume only one technology available to produce a given output, with no scope for substituting other inputs for energy. Others assume technologies will switch in response to small changes in relative input price. Some models assume that once an investment is made, it cannot be altered until its useful lifetime is finished. Still other models assume that capital and labor can be shifted costlessly and instantaneously from one use to another. Clearly, models can either overstate or understate the range and ease of possible substitutions. It is difficult for modelers to get it right.

C. TECHNOLOGICAL CHANGE AND ENERGY EFFICIENCY IMPROVEMENT

Redesigning existing products and processes and introducing new ones can also save energy, but the oversimplified assumptions about technological change in most aggregated models are unrealistic. Most models assume a steady annual percentage improvement in energy efficiency, constant across all industries and over time. This percentage rate of improvement, captured by a variable entitled 'autonomous energy efficiency improvement' (AEEI), strongly influences projected energy consumption and emissions, with or without a carbon tax. Though it has been difficult to reach consensus on a proper value for this parameter,[8] its effect is critical: assuming an annual rate of improvement in energy efficiency of 1 percent rather than 0.5 percent can cut projected 2100 emissions levels by half, markedly affecting the projected costs of meeting a CO_2 emissions target (Gaskins and Weyant, 1993). The higher the AEEI, the lower the projected baseline emissions will be without any carbon tax and the lower the additional emissions reduction that will be required to meet any target.

For example, one model predicts that the present value of the cost of reducing emissions to 20 percent below 1990 levels drops from $1 trillion to a negligible level as AEEI rises from 0.5 percent to 1.5 percent (Manne and Richels, 1990a). With these assumptions, most of the fall in emissions comes from energy efficiency improvements that occur with or without a carbon tax.

More critical than the assumed value for AEEI, though, is the assumption that the pace of energy efficiency improvements is independent of energy price changes and policies. Technological changes do react to market incentives provided either by regulation or by energy prices. Deregulation of electricity markets in the United States has already accelerated market penetration by high-efficiency, low-cost generating technologies. In general, the phenomenon of "induced technological change" is well-recognized (Binswanger and Ruttan, 1978). Economists have shown that rising energy prices induce more rapid rates of energy-saving innovation (Newell et al., 1996; Grubb et al., 1995). Recent modeling analysis in which higher energy prices are assumed to stimulate energy-saving technological change predict lower economic costs of meeting given carbon abatement targets (Goulder and Schneider, 1996).[9]

D. BACKSTOP ENERGY SOURCES

Non-fossil energy sources do exist: hydroelectricity, nuclear power, wind and solar energy, and biomass, to name the most significant. As the prices of fossil fuels rise, these and other alter-

[8] The Clinton administration's Interagency Analytical Team, convened in early 1997 to examine the economic impacts of carbon abatement policies through modeling exercises, settled on a value of 1 percent per year.

[9] If induced technological change is assumed and abatement costs fall in response to a carbon tax, more abatement will occur and *total* (as opposed to per unit) abatement costs will be higher than if technological change is not induced. Then, since total emissions will be lower, the predicted economic impact will depend on whether the economic damages from climate change are taken into account.

Economists have shown that rising energy prices induce more rapid rates of energy-saving innovation.

TABLE 1	COST OF ELECTRIC POWER GENERATION IN THE UNITED STATES, 1985, 1994, 2000		
Technology	**1985**	**1994**	**2000**
	(in 1993 cents per kilowatt-hour)		
Natural Gas	10-13	4-5	3-4
Coal	8-10	5-6	4-5
Wind	10-13	5-7	4-5
Solar Thermal[1]	13-26	8-10	5-6
Nuclear	10-21	10-21	-[2]

[1] With natural gas as a backup fuel

[2] No plant ordered since 1978

Source: Worldwatch Institute, 1994

native energy sources become more attractive. Though too expensive to be widely used today, technological improvements may eventually bring the costs of alternative energy sources down below the costs of fossil fuels in many applications if carbon taxes or comparable policies raise fossil fuel prices. As Table 1 shows, wind and direct solar energy have become more nearly competitive with coal and natural gas sources over the past two decades. Hence, at some point, if carbon emissions are restricted, non-carbon 'backstop' technologies may begin to replace conventional fuels as major energy sources. The concept of a backstop technology was popularized during the oil crises of the 1970s. It refers to an alternative energy source available in virtually unlimited quantities at some price. One example is solar energy.

Modelers have treated non-fossil energy sources in various ways, making different assumptions about their availability, initial cost, and subsequent cost changes. Some models, which have excluded backstop technologies, overstate the economic impacts of a carbon tax because the economy was assumed to rely indefinitely on conventional fuels even after their costs were higher than the costs at which alternative energy sources would be profitable. Accordingly, the carbon tax would have to continue to rise indefinitely to hold carbon emissions constant despite economic growth (Dean and Hoeller, 1992).

Other models have recognized non-fossil energy sources, but assume that their availability is limited so that their prices will rise when used in greater volumes. This assumption is reasonable for sources such as hydroelectricity from dams, since the number of suitable dam sites is limited, but it may not be appropriate for solar energy. Hence, although the early Edmonds-Reilly model predicted an unexceptional tax of $351 per ton of carbon to reduce emissions by 45 percent in 2020, by 2095 the predicted tax for an 88 percent cut below the baseline projection was $2,754 per ton, reducing annual GDP by an estimated 8.8 percent (Barns et al., 1992; Dean and Hoeller, 1992).

Other models assume that backstop energy sources will be available at non-increasing prices, making the key issue how high that price is assumed to be. Models that assume a very high price (e.g., one equivalent to six times today's average fossil fuel prices) make the

> **Merely stabilizing the emissions rate would allow concentrations to continue rising for centuries.**

availability of non-fossil energy sources virtually irrelevant because it would be uneconomic to use them until the distant future, if at all (Charles River Associates, 1997). Models that assume a more reasonable price predict that stabilizing CO_2 concentrations will cost much less. For example, the GLOBAL 2100 model assumes that non-fossil backstop energy sources will become available at future prices as low as twice current electricity prices. In this model, the estimated carbon tax falls from $354 per ton in 2020 to $208 by 2050 and remains at that level (Manne, 1992). As a result, the model predicts gross costs of only 3.1 percent of GDP, instead of 8.8 percent, for the same 88 percent reduction in projected baseline emissions (Dean and Hoeller, 1992).

Furthermore, were it assumed that the costs of alternative energy sources would decline over time with technological improvements and economies from large-scale production—just as fossil energy costs have done—then the predicted costs of stabilizing CO_2 concentrations would also decline over time. Most models, however, fail to make this plausible assumption about the future costs of non-fossil energy sources, even though their costs have fallen substantially in past decades.

E. THE TIME PATH FOR STABILIZING CO_2 CONCENTRATIONS

Ultimately, the goal of climate policy must be to stabilize the atmospheric *concentration* of carbon dioxide and other greenhouse gases rather than the rate of CO_2 *emissions*. These two objectives differ substantially because emissions remain aloft in the atmosphere for substantial periods of time. Merely stabilizing the emissions rate would allow concentrations to continue rising for centuries. However, since national and international policy commitments have been framed in terms of emissions, modelers have adopted the same perspective. Recent analysis shows that adopting an explicit long-term target for atmospheric concentrations and then choosing policies to achieve the most efficient time path for emissions reductions to meet the target could significantly lower the economic impact (Wigley et al., 1996).

A target for atmospheric concentrations of CO_2 is like a 'carbon budget' that limits total CO_2 emissions within a specified period of years. Under some circumstances, it might be cheaper to use up more of the budget early on, postponing cutbacks until later (Richels and Sturm, 1996). Because the capital stock is durable and because so much equipment and building will ultimately have to be changed, adjustment is a costly process. If the time allowed for transition to a lower emissions path is increased to allow capital stock to be replaced as it wears out, overall abatement costs could be reduced. Also, as research and development yields new superior energy-savings technologies, more low-cost substitutes should be available in later years. Finally, postponing costs reduces them because, with a positive return on investment, fewer resources need to be set aside today to meet future costs.

With these assumptions, some models have suggested that it may be more cost-effective to allow emissions to rise for some decades before restricting them significantly below 1990 levels. According to one analysis of the cost-effective pattern of emissions reductions for the period 2000 to 2050, flexibility in the timing of global reductions could lower overall costs by more than 35 percent, compared to a less flexible program that achieved the same atmospheric concentration (Richels et al., 1996). Annual GDP losses would be higher towards the end of the period but would be more than offset by considerably smaller losses in earlier years.

Under other assumptions, however, the least-cost approach would be to avoid the buildup of emissions and consequent steep decline later in the period by instituting carbon taxes earlier on. The key is adopting policies to encourage early development of energy-efficient and low-carbon technologies and to discourage long-lived investments in carbon-intensive energy facilities. According to one recent assessment,

> The window of opportunity for reducing cost implies a need for immediate and continuing action to develop new low-carbon technologies and to begin shifting long-lived investment decisions toward alternatives that lower carbon emissions. Absent these actions, the rapid future emissions reductions included in the delayed emissions scenario may be more costly than more evenly paced, and earlier, reductions (Jaccard and Montgomery, 1996).

Some models assume that policymakers can take advantage of their constituents' foresightedness by simply announcing their intention to limit emissions later in the period without taking any such measures immediately. These models assume that businesses and households will immediately revise their research and investment strategies in response to such an announcement to minimize their exposure to expected future energy price increases. Under these assumptions, the economy can start enjoying the benefits of redirected R&D expenditures and investments without incurring the costs of higher energy carbon taxes.

More realistically, investors may doubt whether a government that declines to institute carbon taxes or equivalent policies to meet its international commitments today will be certain to do so ten or more years into the future. Skeptical investors may adopt a wait-and-see attitude. Should that be the case, equipment and buildings embodying high-carbon technologies will continue to accumulate, making the sharp future reductions needed to stay within a carbon budget all the more expensive. To quell such doubts, sending a credible policy signal by instituting a small but unmistakable policy measure right at the start would be less costly in the long run. A carbon tax that is introduced at a low level and rises—perhaps significantly—in future years would make sense.

Sending a credible policy signal by instituting a small but unmistakable policy measure right at the start would be less costly in the long run.

> **International joint implementation of CO_2 emissions reductions would allow a utility in Norway to achieve an emissions reduction by contracting to pay a factory in Poland to install more fuel-efficient furnaces.**

F. THE POSSIBILITY OF INTERNATIONAL JOINT IMPLEMENTATION

The U.S. Environmental Protection Agency has long recognized that allowing emissions sources to contract with each other to implement required emissions reductions can substantially reduce the costs of achieving any overall abatement target. When able to enter into such transactions, a facility that could reduce emissions only at great cost can compensate another facility with less expensive abatement possibilities to make the cuts for it. This is called "joint implementation." If the emissions from both facilities contribute equally to the ambient pollution problem, joint implementation of an abatement target—or voluntary emissions trading, as it's also called—can be advantageous to both parties and save money. Emissions trading has been used in the United States to reduce the costs of cutting lead, hydrocarbon, and sulfur oxide emissions. The United States has been a leader in developing this policy approach.

International joint implementation of CO_2 emissions reductions would allow a utility in Norway to achieve an emissions reduction by contracting to pay a factory in Poland to install more fuel-efficient furnaces. It would allow a major multinational company to achieve a targeted emissions reduction by energy-saving measures at any of its facilities around the world. Since CO_2 has the same effect on climate wherever it is released, finding the lowest-cost abatement possibilities is cost-effective. Making such trades voluntary would ensure that both parties to the transaction gain from it.

So far, joint implementation of CO_2 abatement has been tried only in experimental pilot programs: U.S. utilities have financed reforestation programs in Central America to capture CO_2 from the atmosphere, for example. Joint implementation cannot be used more widely until countries have set binding emissions reduction targets (Zollinger and Dower, 1996). But, getting countries to agree on the baselines that should apply to each, from which emissions reductions will be measured, is a formidable task. Moreover, monitoring and verification of emissions reductions, along with some mechanism to enforce contractual obligations, will be essential if joint implementation is to work. Nonetheless, such implementation offers a promising opportunity to lower CO_2 abatement costs and the economic impacts of protecting the climate. Models focussed only on the U.S. economy generally assume this possibility away. Efforts to model other parts of the world along with the United States have explored the potential savings from joint implementation and have found them to be substantial. [10]

G. HOW REVENUES ARE RECYCLED INTO THE ECONOMY

A carbon tax would generate significant tax revenues—up to $300bn per year by 2020 for a policy that holds

10 International emissions trading can be modeled as if all countries set the same carbon tax rate, so that cost-effective emissions reductions are advantageous to undertake in whatever country they arise.

carbon emissions to 80 percent of their 1990 levels (Gaskins and Weyant, 1993). Starting from a position of fiscal balance, some mechanism for recycling these revenues into the economy is needed to prevent a general deflationary impact. Assuming that revenues would not be fully recycled was one of the reasons for an early pessimistic conclusion that a carbon tax of $100 per ton phased in over 10 years would reduce GDP by 2 percent annually and cut projected baseline emissions by only a small amount (CBO, 1990).

Economic modelers then began assuming that revenues would be returned in the form of lump-sum rebates. (A lump-sum tax or rebate is one in which the amount transferred is completely independent of all taxpayer behavior. Increasing the personal exemption in the income tax would be an example of a lump-sum rebate.) By assuming neutral lump-sum recycling, modelers tried to separate the economic impacts arising from climate abatement policy from those arising from other tax cuts (Gaskins and Weyant, 1993).

However, the baseline projections from which economic impacts are estimated implicitly or explicitly recognize that the U.S. economy's tax structure reduces private incomes by far more than a dollar for every dollar of tax revenue collected. The disincentive effects of taxes on payrolls, incomes, and profits either reduces work, investment, and savings or diverts them into less productive forms to reduce tax liabilities. Economists pointed out long ago that using carbon tax revenues to reduce existing taxes that penalize work, savings, and investment would lower the net cost of reducing emissions (Terkla, 1984). Indeed, this tax-shifting opportunity is one of the principal reasons for favoring a carbon tax (or carbon permits that are auctioned off) over alternative climate policy instruments.

Taxes on labor earnings drive a 'wedge' between the price paid by firms for labor and the wage received by the employee. As a consequence, fewer workers are employed at a lower real wage rate, and the economy suffers accordingly (MaCurdy et al., 1990; Triest, 1990; Browning, 1976). Similarly, taxes on investment earnings reduce the amounts saved, resulting in lower rates of capital investment and economic growth. Empirical work suggests that these efficiency losses may be large: for every dollar of revenue raised from taxes on labor earnings, private income might ultimately fall by $1.10 to $1.25; and, per dollar of revenue from taxes on investment earnings, income might fall by $1.50 to $1.95 (Nordhaus, 1993; Fullerton, 1991; Jorgenson and Yun, 1990; Ballard, Shoven and Whalley, 1985).

The use of carbon tax revenues to reduce existing tax rates benefits society not only by correcting a market failure in energy use but also by reducing the costs of other distorting taxes, a so-called "double dividend" (Repetto et al., 1992). If the efficiency benefit from reducing existing taxes on labor and capital is incorporated into the analysis, the projected net economic

Using carbon tax revenues to reduce existing taxes that penalize work, savings, and investment would lower the net cost of reducing emissions.

TABLE 2 GDP LOSS (1990-2010) UNDER ALTERNATIVE RECYCLING OPTIONS

Recycling Option	Model			
	DRI	LINK	DGEM	Goulder
	(Percentage change in discounted constant price GDP)			
Lump Sum Tax Cuts	-0.58	-0.46	-0.62	-0.24
Personal Income Tax Cuts	-0.56	-0.53	-0.16	-0.16
Corporate Income Tax Cuts	0.40	-0.11	0.60	-0.17
Payroll Tax Cuts				-0.18
Employee Only	-0.58	-0.53		
Employer Only	0.19	-0.25		
Investment Tax Credit	1.55	1.67		0.00

Source: Shackleton et al. (1992).

impacts of a carbon tax are substantially more favorable than they appear to be when lump-sum revenue recycling is assumed (Goulder, 1995).

One analysis of this issue compared six revenue-recycling options, using four different models (Shackleton et al., 1992). Despite modeling differences a clear hierarchy of recycling options emerges, with lump-sum rebates among the least attractive. As Table 2 shows, using revenues to lower corporate income tax and payroll taxes instead of giving lump-sum rebates significantly reduces the adverse economic impacts of a carbon tax. Not surprisingly, the analysis reveals that recycling revenues by reducing taxes on capital investment earnings and thereby encouraging capital formation reduces the economic impacts more than lowering taxes on consumption. In two models, recycling revenues by introducing an investment tax credit leads to net gains in GDP and consumption. (In a third model, the same recycling mechanism exactly offsets the costs of mitigation.) More recent studies support these findings (McKibbin and Wilcoxen, 1996; Goulder, 1995; Jorgenson et al., 1995).

Models suggesting that the substitution of a carbon tax for other taxes could provide net economic benefits, *irrespective of the environmental gains*, prompted some economists to propose a stronger form of the double dividend hypothesis, claiming that a carbon tax could be justified even apart from environmental benefits. This has led to a debate about whether the revenue recycling effect alone is sufficient to justify the levying of a carbon tax (Goulder, 1995; Parry, 1995; Bovenberg and de Mooij, 1994). This is obviously not always true. If the environmental benefits and the welfare gain from correcting a market failure are ignored, a tax on fuels distorts energy markets just like any excise tax does. It may be more or less distortionary than another tax already being levied, depending on the size of energy markets, the price sensitivity of energy demands, the structure of pre-existing energy taxes, and other conditions. Though Table 2 shows particular tax substitutions that suggest a strong double dividend, economists doubt whether a strong double dividend would generally be available. The key issue is how an energy tax would interact with existing taxes on capital

BOX 2 TAX INTERACTIONS AND THE DOUBLE DIVIDEND DEBATE

Several economists have pointed out that when distorting taxes on labor and capital income already exist in an economy, the effects of replacing some of these taxes with a carbon tax are complicated. A carbon tax, which raises the prices of energy and energy-intensive goods, would raise consumers' cost of living and,with wage and salary levels unchanged,would lower real wages. In this respect, a carbon tax would have much the same effect as a direct tax on labor incomes. By lowering the real rewards from working, both kinds of tax would distort labor markets and reduce labor supply. Substituting one for another might not reduce the labor market distortion. Indeed, because a carbon tax—with a narrower tax base—would require a higher tax rate than a direct payroll tax to raise the same revenue, a revenue-neutral tax shift might even reduce labor supply further. Using this reasoning, these economists have questioned the likelihood of a strong double dividend and the extent of a weak double dividend.

However, this argument rests on special assumptions. It assumes unrealistically that a carbon tax would be borne entirely by working consumers, not at all by owners of coal mines, stockholders in energy companies, or foreign oil sheikhs. Clearly, this assumption is not shared by these parties themselves, as they vigorously oppose the imposition of a carbon tax or any equivalent policy. To the extent that a carbon tax would not reduce the real value of labor incomes but would fall instead on earnings from capital and property or on foreign suppliers, it would not have the effect of reducing labor supply.

More importantly, this line of reasoning ignores the effects of environmental damages on the cost of living. A carbon tax would raise energy prices but reduce CO_2 emissions and other pollutants stemming from fossil fuel combustion: sulfur and nitrogen oxides, hydrocarbons and other smog precursors, particulates, and carbon monoxide. Cutting back emissions of these pollutants would decrease medical expenses and the number of work days lost to sickness. For example, a recent comprehensive study estimated that improved air quality under the Clean Air Act has resulted in striking health benefits to the American population, including 80,000 fewer heart attacks in 1990, 10,000 fewer strokes, 15,000 fewer cases of respiratory illness, 13,000 fewer cases of hypertension, and a long list of other health benefits (U.S. EPA, 1996). These improvements are much more than an additional fuel tax would generate, but make the point that health benefits from reducing air pollution are substantial. Improved health and reduced health expenditures stemming from a carbon tax would offset the tax's modest negative impact on incentives to work.

Similarly, all assessments of the risks from climate change emphasize that global warming may raise production *costs* in affected industries. Increased evapotranspiration may reduce crop yields and water availability, raising agricultural production costs. Higher temperatures may raise air conditioning and cooling costs (though lowering heating bills). Rising sea levels may increase coastal flooding and storm damages. Warmer climates may increase the range of agricultural pests and disease vectors. Production losses from a doubling of CO_2 concentrations are estimated at 1 or 2 percent of GDP (Fankhauser, 1995; Tol, 1995; Cline, 1992). Again, these estimated damages are larger than those that a modest carbon tax would eliminate, but make the point that forestalling climate change would reduce many production costs. Therefore, though a carbon tax would raise energy costs it would *lower* costs in a range of other industries, with uncertain effects on the overall cost of living. On balance, taking both costs and benefits into account, the interaction of a carbon tax or equivalent policy with pre-existing taxes is as likely to magnify the double dividend as to shrink it.

and labor. *(See Box 2.)* Nonetheless, models that ignore the potential gains from substituting a market-correcting tax for a market-distorting tax are excessively pessimistic about the economic impacts of a carbon tax.

H. AVOIDING THE COSTS OF CLIMATE CHANGE

Most models used to simulate the economic impacts of carbon taxes have been adaptations of existing models designed for energy planning or for

Heat waves, severe storms, and other extreme weather episodes may become more frequent as average temperature and precipitation increase.

general macroeconomic forecasting, and were not able to incorporate the benefits of preventing climate change, or equivalently, the costs of doing nothing. Instead, they have simply measured the economic impacts of meeting emissions abatement targets, implicitly assuming in the baseline projections that climate change would have no effect on the economy.

Granted, predicting the physical effects of global warming is difficult enough, let alone estimating the effect on human welfare. Climate change is likely to have many economic effects, some favorable and others not, especially on agriculture, fishing, forestry, recreation, flood control, water supply, and other industries dependent on natural resources. Other impacts will occur not through market mechanisms but through ecological changes, altered rates of human morbidity and mortality, heat-related distress and discomfort, and migration (IPCC, 1996b).

Several studies have estimated damages from a doubling of CO_2 concentrations midway through the next century. [11] Available estimates rely on questionable simplifying assumptions and are little more than best guesses, expressed as wide ranges. Estimates for the U.S. economy range from 1.0 to 1.5 percent of GDP annually (Fankhauser, 1995; Tol, 1995; Cline, 1992; Nordhaus, 1991). Titus (1992) predicts annual damages of 2.5 percent of GDP, largely because he assumes that CO_2 doubling will lead to higher temperatures than other researchers do. Damages for the global economy as a whole are closer to 2 percent of world GDP annually and up to 9 percent of national GDP for some developing countries (Fankhauser, 1995; Tol, 1995; Fankhauser and Pearce, 1994).

These estimates might be too pessimistic (Fankhauser and Tol, Energy Policy, 1996). They do not adequately reflect the possibility that adaptations to gradual climate change, such as changed agricultural practices, may be available to forestall damages at relatively low cost (Mendelsohn, 1996; Adams et al., 1994; Nordhaus, 1994). They may give too little weight to potential cost savings from climate change, such as reduced heating costs (Rosenthal et al., 1995). And they may not adequately reflect changing scientific perceptions of likely ecological changes, such as lowered predictions of sea level rise (Yohe et al., 1996).

On the other hand, these estimates refer only to CO_2 doubling, when much higher concentration levels are likely in the more distant future. Even if emissions are successfully stabilized and held at 1990 levels, concentrations will double by 2100 and increase subsequently thereafter (IPCC, 1996a). Moreover, if emissions continue rising, concentrations could exceed 700 ppmv by the end of the next century, and damages may increase more than proportionately (IPCC 1996a; Cline, 1992).

In addition, estimates typically are based on increases in average temperature and precipitation, but heat waves, severe storms, and other

11 'Doubling' refers to CO_2 concentrations of 560 ppmv (parts per million by volume), twice the pre-industrial level of 280 ppmv that forms the standard benchmark. IPCC (1996a) estimates the temperature rise from doubling at 1-3.5C with a 'best guess' of 2.5C—the figure adopted by most researchers.

extreme weather episodes may become more frequent as average temperature and precipitation increase. Omitting these extremes probably understates the impacts of climate change, since gradual changes in average temperature and rainfall have much less effect than the occasional extreme event does. The ultimate manifestation of this is the unlikely possibility of catastrophic events such as the disintegration of the West Antarctic ice sheet, the redirection of the Gulf Stream, the spread of disease vectors into unresistant populations, or the possibility of a 'runaway' greenhouse effect, where positive feedbacks overwhelm negative ones (IPCC, 1996b). Potential catastrophes seem more likely the longer the time frame under consideration and cannot be assigned a zero probability.

Uncertain though they are, there are costs associated with doing nothing in the face of rising greenhouse gas concentrations. These are typically not reflected in the baseline projections of most models. Therefore, the net economic impacts of carbon taxes tend to be overstated. The few models that do take expected damages from climate change into account predict that a carbon tax set at an appropriate rate, with revenues recycled efficiently back into the economy, actually improves economic welfare (Nordhaus and Young, 1996; Jorgenson et al., 1995; Nordhaus, 1994, 1993).

I. AVOIDING OTHER ENVIRONMENTAL DAMAGES

By discouraging coal and other fossil fuel consumption, a carbon tax would reduce emissions of such air pollutants as carbon monoxide, sulfur and nitrogen oxides, and toxic trace pollutants in exhaust gases. Though most of these pollutants are already regulated, this usually takes the form of a maximum rate of emissions per million BTUs. Thus, despite regulations, reducing the number of BTUs generated by fossil fuel combustion would lower emissions. This would almost immediately reduce damages to health, visibility, materials, and crops.

A simple approach to assessing non-climate benefits is to calculate the tax-induced reduction in fossil fuel consumption and use emissions profiles for individual fuels to estimate overall emissions reductions (Scheraga and Leary, 1993). This approach suggests that a carbon tax set to keep CO_2 emissions at their 1990 levels by the year 2000 would reduce various atmospheric emissions below projected baselines by 1 to 7 percent. EPA estimates that the economic savings in the year 2000 from the resulting pollution abatement would be between $300 million and $3 billion.

Other studies in Europe and the United States estimate that the non-climate benefits of a carbon tax would probably be as large or larger than the benefits of avoiding climate change (Ekins, 1995; Jorgenson et al., 1995).

Studies in Europe and the United States estimate that the non-climate benefits of a carbon tax would probably be as large or larger than the benefits of avoiding climate change.

These estimated economic savings from reduced air pollution damages may be sufficiently large to offset from 30 to 100 percent of carbon abatement costs (IPCC, 1996b).

Large as these sums are, this approach underestimates damages over time because it ignores the fact that in baseline projections, in the absence of carbon taxes, rising population and economic growth would increase exposure to air pollution while increased fuel consumption would reduce air quality. Current environmental standards typically limit emissions on a per-BTU basis for stationary sources and on a per-mile basis for vehicles. As the economy grows, trying to maintain environmental quality by making each mile traveled or BTU generated less polluting can become increasingly costly. At some point, fuel taxes that provide incentives to conserve energy and limit the amount of driving become cost-effective adjuncts to such regulations (Eskeland and Devarajan, 1996).

One of the few models to incorporate the savings from reducing pollution damages estimates how much reduction in CO_2 emissions could be had for "free" under a "no regrets" policy—a policy under which net abatement costs are exactly offset by non-climate benefits. When firms are assumed to be able to substitute labor and capital for energy, a reduction of nearly 50 percent of baseline emissions can be attained with no loss in economic welfare (Boyd, Krutilla and Viscusi, 1995).

Despite these impressive magnitudes, most models of carbon abatement policies have ignored the potential savings in pollution damages, and the remaining few have incorporated them rather crudely. More recent and comprehensive research on the benefits of air pollution control is now available to be incorporated into climate models (U.S. EPA, 1996). Other benefits from lower energy consumption, such as reduced road congestion and fewer environmental impacts from fuel mining, refining, and transport, could also be taken into account.

TRADE AND EQUITY ISSUES

4

A. TRADE ISSUES

Concerns about adverse international trade consequences are strong among businessmen who fear that if the United States unilaterally adopted a carbon tax its energy-intensive industries almost certainly would become less competitive in international trade. International shifts in the location of production, so-called 'leakage' effects, would mean that cuts in domestic emissions would be partially offset by increases in emissions abroad as energy-intensive industries expanded overseas (Barrett, 1994; Gaskins and Weyant, 1993). For these reasons, a carbon tax adopted unilaterally would be more costly to the U.S. economy and less effective in reducing global emissions than one adopted by many countries acting together.

Empirical research into the magnitude of leakage effects is somewhat inconclusive. The penalty for acting unilaterally would be greater for small economies than for large economies like that of the United States. Studies of unilateral emissions reduction policies in OECD countries predict leakage rates of between 3.5 and 70 percent [12] (Manne, 1993; Oliveira-Martins et al., 1992; Pezzey, 1992; Rutherford, 1992). Studies in the United States suggest that cost differences created by international differences in environmental policies have minor impacts on trade and investment flows (Repetto, 1995).

There would also be international trade gains, however. Given the size of the U.S. market, a tax-induced fall in U.S. energy consumption would help restrain world fuel prices, improving our terms of trade as a net oil-importing economy. Baseline projections suggest that in the absence of higher energy taxes, fossil fuel consumption in the United States and the rest of the world will continue to grow substantially in coming decades. One implication, given the already declining oil production in this country, is that OPEC will supply an increasing share of world petroleum markets, and the United States will become increasingly dependent on imports from OPEC producers. OPEC's world market share might reach 72 percent by 2015, more than in 1974 when its supply restrictions precipitated the first "oil shock" (U.S. EIA, 1996). A carbon tax would mitigate these energy security and trade risks.

If the United States adopted a carbon tax unilaterally, the risks of climate change faced by other countries would fall, reducing their incentive to act (Barrett, 1994). However, increasing international coordination within the Framework Convention on Climate Change means that unilateral action is unlikely. European countries have stated their willingness to adopt a carbon tax if major trading partners—the United States in particular—do the same (Commission of the European Communities, 1992). Sweden and Norway already have signaled willingness by adopting small taxes. Developing countries are clearly unwilling to commit themselves to action unless industrial countries that have emitted most of the carbon dioxide to date take the lead. Nonetheless, substantial economic

12 The leakage rate is measured as the increase in emissions outside the tax region over the emissions reduction within the region.

reforms and restructuring of energy markets have taken place in developing and transitional economies, for purely economic reasons. Energy subsidies have been reduced, subsidies to energy-intensive industries have fallen, and institutional changes have opened the energy sector to improved technology and greater efficiency. These significant changes have already reduced CO_2 emissions in these countries well below their projected levels (Reid and Goldemberg, 1997). Though undertaken for strictly economic reasons, these policy changes in other countries have greatly reduced the risks of unilateral U.S. action.

B. DISTRIBUTIONAL EFFECTS

Even if the macroeconomic economic impacts of a carbon tax were small or positive, some household, regional and sectoral groups might be adversely affected, if they are highly dependent on fossil fuels and have limited ability to adjust to price changes. These distributional considerations are important policy issues: a carbon tax program may be considered undesirable if costs are felt disproportionately by low-income groups or specific industry sectors, even if predicted overall economic burdens are small.

1. DISTRIBUTIONAL EFFECTS ON HOUSEHOLDS

It is commonly reckoned that a carbon tax would be regressive, hitting lower-income households proportionately harder than those better off. To illustrate, a $100 per ton carbon tax instituted in 1990 would have created a burden amounting to 10.1 percent of income for the lowest decile of households arrayed by household income but only 1.5 percent for the top decile (Poterba, 1991). However, when households are arrayed by expenditure instead of income the carbon tax is much less regressive, accounting for 3.7 percent to 2.3 percent of total expenditures for the bottom and top deciles respectively. Household expenditure is a better indicator of long-term ability to pay—and hence a better base against which to judge distributional effects—because it is more stable over time than household income. Short-term fluctuations in income are typically cushioned by borrowing or adding to savings. At any time, there will be many households whose incomes are temporarily and abnormally inflated or depressed.

At best, such calculations provide imperfect guides: they reflect only the direct effect of higher energy prices on expenditures, omitting the fact that goods whose production relies on fuel—everything from cars to foodstuffs—will exhibit price rises too. A Canadian study that accounted for direct and indirect effects by assessing the carbon-intensity of the complete set of purchases made by different income or expenditure groups found that a $100 carbon tax would only be mildly regressive even for households grouped by income. The burden was 2.9 percent for the bottom income quintile and 1.8 percent for the top quintile (Hamilton and Cameron, 1994).

However, since the long-run distributional impact will also depend on

European countries have stated their willingness to adopt a carbon tax if major trading partners—the United States in particular—do the same.

The energy industry generally, and the coal mining industry particularly, will experience the greatest impacts from a carbon tax.

changes in employment, wages, and asset prices, only overall economic models can make comprehensive estimates. An analysis using a general equilibrium model traced the welfare impacts of a carbon tax on 16,128 separate household groups categorized by family size, age and sex of household head, race, region, type of residence, and wealth. The model predicted that, on the whole, the distributional effects of a carbon tax would be modest, ranging from mildly progressive to mildly regressive, depending on underlying assumptions (Jorgenson et al., 1992).

Of course, the overall distributional effect of a carbon tax would depend on the use of the revenues. Expanding the earned-income tax credit would benefit low-income households; reducing capital gains taxes would benefit the rich. There are trade-offs between equity and efficiency concerns. Efficiency gains seem greatest if taxes on investment income are reduced; equity objectives are better served if taxes on labor earnings are reduced. A combination of the two—using the revenues to cut both capital and labor taxes—would balance these objectives.

Moreover, the distributional impacts of a carbon tax should be judged against the impacts of a business-as-usual policy. A number of studies predict that global warming would lead to higher food prices. Since food is as much a necessity as energy is, a business-as-usual policy might also be regressive (Scheraga et al., 1993). Similarly, since poorer households typically inhabit areas of lower environmental quality and are generally in poorer health, they benefit disproportionately from an improvement in air quality (Yin, 1993). However, they may value these relatively large benefits absolutely less than the smaller benefits felt by better off households (Harrison and Rubinfeld, 1978). All in all, the effects of a carbon tax on the distribution of income and expenditures among households would probably be small; and effects on vulnerable groups could rather readily be offset by cost-of-living adjustments in government transfer programs and by other policy instruments.

2. DISTRIBUTIONAL EFFECTS BY SECTOR AND REGION

The energy industry generally, and the coal mining industry particularly, will experience the greatest impacts from a carbon tax, under which the tax rate on coal, a carbon-intensive fuel, will be relatively high. Compared to the business-as-usual scenario, coal output may decrease by 25 percent and prices may rise by 35 percent by 2020 under a tax that stabilizes emissions at 1990 levels (Jorgenson et al., 1992). Electric utilities, the next most affected sector, face output losses and price increases of just over 5 percent compared with the projected baseline, mainly because they use coal. But these reductions do not imply absolute declines because substantial growth in output is projected for both sectors in the baseline. Some railroads would suffer adverse impacts since coal makes up a substantial fraction of their freight traffic. Of course, other industries, such as the natural gas and renewable energy industries, would probably do better if a carbon tax were enacted.

Regional impacts of a carbon tax inevitably correspond to sectoral impacts. Coal mining states, such as West Virginia and Wyoming, would be disproportionately affected. But, according to one study, the Pacific Northwest would fare relatively well because of the availability of cheap hydropower—assuming such power continues to be heavily subsidized (DeWitt, Dowlatabadi, and Kopp, 1991). Global warming, like a carbon tax, would also affect some regions and industries disproportionately. Along with coastal areas, regions dependent on agriculture, water supply, and forestry would suffer most (OECD, 1996).

To summarize, studies predict that a carbon tax will be only mildly regressive for households, and those effects could be offset by revenue recycling and cost-of-living adjustments in government transfer programs. The distribution impacts of a carbon tax might well be no worse than those of global warming itself. However, coal mining and coal mining regions would be significantly affected, since coal output would fall relative to a business-as-usual scenario and, assuming a sufficiently high tax, would fall absolutely as well.

CONCLUSIONS AND RECOMMENDATIONS

5

Despite the wide range of predicted economic impacts of CO_2 abatement, two areas of policy agreement among the underlying economic models stand out. First, the economic impacts will be much more favorable if revenue-raising policy instruments are used to control CO_2 emissions and the revenues are used to reduce other burdensome tax rates. As explained above, the two principal revenue-raising policy instruments are carbon taxes and tradable permits that are auctioned off by the government rather than given away. Both instruments imply that carbon-based fuels will become more expensive to the users, raising costs throughout the economy. However, using the revenues to lower taxes on labor and capital would offset some of these higher costs and improve the economic impacts.

Giving permits away to energy users and sellers or using regulations to force industries to reduce their CO_2 emissions are alternative policy approaches that forgo potential revenues. These approaches would also make fossil fuels more expensive to the users, raising costs throughout the economy. But, they would have quite different economic implications because they create no opportunities to generate offsetting efficiency gains by reducing the marginal rates of distorting taxes.

Instead, the potential revenues from auctioning off permits to burn fossil fuels remain with the firms that received free permits or regulatory permission. Conceding these permits to firms may be politically advantageous because it awards a valuable right to certain firms and industries that might otherwise oppose any climate protection policy. However, economic models agree that this political advantage would be bought at a high economic price. Income and economic growth would be lower if this approach were adopted, just as they would be if revenues from a carbon tax were recycled through lump-sum rebates.

Similarly, forgoing carbon tax revenues would limit the government's opportunities to finance offsetting tax cuts and other programs to cushion the distributive burden of higher energy prices on lower-income households and other vulnerable groups. Instead, those companies permitted to sell or use the limited quantities of carbon fuels available would obtain windfall profits. For both equity and efficiency reasons, the U.S. Government should base its climate

Climate protection plans should be revenue-neutral. They should include explicit proposals for offsetting cuts in distorting taxes on labor and investment incomes.

protection plans on policy instruments such as a carbon tax or auctioned-off carbon permits that yield revenues to finance tax reductions. Moreover, climate protection plans should be revenue-neutral. They should include explicit proposals for offsetting cuts in distorting taxes on labor and investment incomes. Identifying these tax cuts in advance and ensuring that the government's spending and deficit are unchanged should help allay economic concerns about climate protection.

The second area of agreement among economic models that have examined the issue is that joint implementation will reduce the overall costs of achieving carbon abatement targets. Emissions trading among domestic U.S. firms and facilities has already proven to be cost-effective in reducing conventional air pollutants. The gains from international joint implementation of CO_2 abatement programs should be even larger because the costs of reducing fossil fuel use vary so widely among nations. International joint implementation is compatible both with a U.S. carbon tax and with a carbon permit trading system. The United States has consistently favored joint implementation in principle as a cost-effective market-based policy approach. Ironically, many countries with emerging market economies have not approved the principle of such implementation even though they would benefit significantly from the infusion of technology and investment that it would bring. The U.S. Government should continue to consult and negotiate with other nations to gain international acceptance for joint implementation among nations adopting CO_2 reduction commitments, and to promote joint implementation through pilot projects and institutional development.

Aside from the effects of these two policy choices, the remaining variation in the predicted economic impacts of climate protection stems from differences in underlying assumptions built into economic simulation models. Predictions that a carbon tax or a cap-and-trade policy to reduce CO_2 emissions would seriously harm the economy are unrealistic. They stem from worst-case modeling assumptions. Under more reasonable assumptions and preferable policy approaches, a carbon tax is a cost-effective way of reducing the risks of climate change and would do no damage to the economy. More likely, taking the environmental effects into account, it would bring long-term benefits.

This point is likely to be obscured as the debate over U.S. climate policy becomes increasingly intense and politicized as international negotiations proceed. Various interests will tout their preferred predictions and experts. The media, on the principle that controversy generates interest, will highlight the divergent predictions rather than informing the public about the underlying sources of disagreement. The public may become confused, disillusioned, or indifferent.

To counter this, it is important that when the administration finalizes its

Many emerging market economies have not approved the principle of joint implementation even though they would benefit significantly from the infusion of technology and investment that it would bring.

proposed climate policy it issues a public statement incorporating its best predictions of the impacts these proposed measures would have on the economy, making clear the assumptions and models on which these estimates are based. When Congress holds televised hearings on this issue, expert witnesses and administration officials should be questioned closely on the assumptions underlying their predictions regarding the effects of climate protection on the economy. The press and other media can contribute significantly to public understanding and debate of this important policy issue by in-depth reporting and analysis that goes beyond the formulaic juxtaposition of opposing viewpoints.

The real issues that need resolution are how to cushion the impacts on those few industries, regions, and communities that would be adversely affected and how to negotiate international agreements that will bring about coordinated actions by all major economic powers. In addition, an overall tax restructuring agenda is needed that would incorporate a phased-in carbon tax or auctioned-off tradable carbon permits and equivalent reductions in other taxes to achieve equity and economic objectives.

REFERENCES

Adams, R.M. et al., (1994), *The Economic Effects of Climate Change on U.S. Agriculture*, Electric Power Research Institute, Palo Alto, CA.

AES Corporation (1990), "An Overview of the Fossil2 Model," Arlington, VA.

Alliance to Save Energy (1993), *Federal Energy Subsidies: Energy, Environmental and Fiscal Impacts*, Washington, D.C.: Alliance to Save Energy.

AOSIS (Alliance of Small Island States) (1995), Draft Protocol submitted to the United Nations Framework Convention on Climate Change, in 'Matters Relating to Commitments,' FCCC/CP/1995/Misc.1.

Ballard C. L., J-B Shoven and J. Whalley (1985), "General Equilibrium Computations of the Marginal Welfare Costs of Taxes in the United States," American Economic Review, 75 (1), pp. 128-138.

Barns, D.W., J. Edmonds and J. Reilly (1992), "Use of the Edmonds-Reilly Model to Model Energy-related Greenhouse Gas Emissions," OECD Dept. of Economic Statistics, Resource Allocation Division, Working Paper No. 113, OECD/GD (91), Paris.

Barrett, S. (1994), "Climate Change Policy and International Trade." In *Climate Change: Policy Instruments and Their Implications, Proceedings of the Tsukuba Workshop of Working Group III*, A. Amaro, B. Fisher, M. Kuroda, T. Morita, and S. Nishioka, *eds*. pp. 15-33, Geneva: IPCC.

Binswanger, Hans P. and Vernon W. Ruttan (1978), *Induced Innovation: Technology, Institutions and Development*. Baltimore: The Johns Hopkins University Press.

Boero, G., R. Clarke, and L.A. Winters (1991), *The Macroeconomic Consequences of Controlling Greenhouse Gases: A Survey*, London: HMSO.

Bovenberg, A. Lans and R.A. de Mooij, (1994), "Environmental Levies and Distortionary Taxation," American Economic Review, September, Vol. 84, no. 4, pp. 1085-1089.

Boyd R., K. Krutilla and W.K. Viscusi (1995), "Energy Taxation as a Policy Instrument to Reduce CO_2 Emissions: A Net Benefit Analysis," Journal of Environmental Economics and Management, Vol. 29, pp. 1-24.

Browning, E.K. (1976), "The Marginal Cost of Public Funds," Journal of Political Economy, 84, pp. 283-298.

Carraro C. and D. Siniscalco (1996), *Environmental Fiscal Reform and Unemployment*, Dordrecht, The Netherlands: Kluwer Academic Publishers.

Charles River Associates (1997), "World Economic Impacts of US Commitments to Medium Term Carbon Emissions Limits," Paper prepared for American Petroleum Institute, CRA No. 837-06.

Cline, W.R. (1992), *The Economics of Global Warming*, Washington D.C., Institute for International Economics.

Commission of the European Communities (1992), "Proposal for a Council Directive Introducing a Tax on Carbon Dioxide Emissions and Energy," Official Journal of the European Community, No. C196, 3.8.92 (August 3rd).

Congressional Budget Office (1990), *Carbon Charges as a Response to Global Warming: the Effects of Taxing Fossil Fuels*, Washington D.C.: Congressional Budget Office (August).

Dean, A. and P. Hoeller, (1992), "Costs of Reducing CO_2 Emissions: Evidence from Six Global Models," OECD Economic Studies, 19 (Winter), pp. 15-47.

De Canio, S. (1993), "Barriers Within Firms to Energy Efficiency Investments," Energy Policy, 21(9), 901-914.

DeWitt D., H. Dowlatabadi, and R. Kopp (1991), "Who Bears the Burden of Energy Taxes?" Discussion Paper QE91-12, Resources for the Future, Washington D.C.

Economists' Statement on Climate Change (1987), Distributed by Redefining Progress, San Francisco, CA.

Ekins, Paul (1995), "Rethinking the Costs Related to Global Warming: A Survey of the Issues," Environmental and Resource Economics, Vol. 6, No. 3, October, pp. 231-277.

Eskeland, G. and Devarajan, S. (1996), "Taxing Bads by Taxing Goods: Pollution Control with Presumptive Charges," Directions in Development, The World Bank.

Fankhauser, S. (1995), *Valuing Climate Change: The Economics of the Greenhouse*, London: Earthscan.

Fankhauser, S. and R. Tol (1996), "Climate Change Costs," Energy Policy, Vol. 24, no. 7, pp. 665-673.

Fankhauser, S. and D.W. Pearce (1994), "The Social Costs of Greenhouse Gas Emissions." In *The Economics of Climate Change: Proceedings of an OECD/IEA Conference*, Paris, Organisation for Economic Cooperation and Development.

Fullerton, D. (1991), "Reconciling Recent Estimates of the Marginal Welfare Cost of Taxation," American Economic Review, Vol. 81, pp. 302-8.

Gaskins D.W. and J.P. Weyant (1993), "Model Comparisons of the Costs of Reducing CO_2 Emissions," American Economic Review (AEA Papers and Proceedings), Vol. 83 (2), pp. 318-323.

Goulder, L.H. (1995), "Effects of Carbon Taxes in an Economy with Prior Tax Distortions: An Intertemporal General Equilibrium Analysis," Journal of Environmental Economics and Management, 29 (3), November, pp. 271-297.

Goulder, L.H. and Stephen S. Schneider (1996), "Induced Technological Change, Crowding Out, and the Attractiveness of CO_2 Emissions Abatement," Working Paper, Stanford University, CA.

Grubb, Michael, T. Chapuis and M. Ha Duong (1995), "The Economics of Changing Course—Implications of Adaptability and Inertia for Optimal Climate Policy," Energy Policy, Vol.23, No. 4/5, pp. 417-432.

Hamilton, K. and G. Cameron (1994), "Simulating the Distributional Effects of a Canadian Carbon Tax," Canadian Public Policy, 20(4), pp. 385-399.

Harrison, D. and D. Rubinfeld (1978), "The Distribution of Benefits from Improvements in Urban Air Quality," Journal of Environmental Economics and Management, Vol 5, December.

Hourcade, J-C and J. Robinson (1996), "Mitigating Factors—Assessing the Costs of Reducing GHG Emissions," Energy Policy, Vol. 24, Nos. 10/11, pp. 863-873.

IPCC (Intergovernmental Panel on Climate Change), (1996a), *The Science of Climate Change*, Vol. 1 of Climate Change 1995: IPCC Second Assessment Report, Cambridge University Press, Cambridge.

IPCC (Intergovernmental Panel on Climate Change), (1996b), *The Economic and Social Dimensions of Climate Change*, Vol. 3 of Climate Change 1995: IPCC Second Assessment Report, Cambridge University Press, Cambridge.

Jaccard, M. and W.D. Montgomery (1996), "Costs of Reducing Greenhouse Gas Emissions in the USA and Canada," Energy Policy, Vol. 24, Nos. 10/11, pp. 889-898.

Jorgenson, D., R. Goettle, D. Gaynor, P. Wilcoxen, and D. Slesnick (1995), "Social Cost Energy Pricing, Tax Recycling and Economic Change," Final Report submitted to the U.S. EPA, EPA contract No. 68-W2-0018, August.

Jorgenson, D., D. Slesnick and P. Wilcoxen (1992), "Carbon Taxes and Economic Welfare," Brookings Papers on Economic Activity: Microeconomics, pp. 432-443.

Jorgenson, D. and K-Y. Yun (1990), "The Excess Burden of Taxation in the U.S.," Harvard Institute of Economic Research Discussion Paper No. 1528, November, Harvard University, Cambridge, MA.

MaCurdy T., D. Green and H. Paarsch (1990), "Assessing Empirical Approaches for Analyzing Taxes and Labor Supply," The Journal of Human Resources, Vol. 25 (3), pp. 415-516.

Manne, A.S. (1993), "International Trade: The Impacts of Unilateral Carbon Emission Limits," paper presented at the OECD International Conference on the Economics of Climate Change, Paris, 14-16 June.

Manne A.S. and R.G. Richels (1990a), "The Costs of Reducing U.S. CO_2 Emission: Further Sensitivity Analyses," The Energy Journal, Vol. 11, no. 4, pp. 69-78.

Manne, A.S., and R.G. Richels (1992), *Buying Greenhouse Insurance*, MIT Press: Cambridge, MA.

McKibbin, W. and P.J. Wilcoxen (1996), "Environmental Policy, Capital Flows and International Trade," Working Paper, Washington D.C.: The Brookings Institution.

Mendelsohn, Robert et al. (1996), "The Economic Impacts of Climate Change in the U.S.," Paper presented at the IPIECA Symposium, Paris, 8-11 October.

National Academy of Sciences (1991), *Policy Implications of Greenhouse Warming*, Washington D.C.: National Academy Press.

Newell, Richard G., A. Jaffe and R. Stavins (1996), "Environmental Policy and Technological Change: The Effect of Economic Incentives and Direct Regulation of Energy-Saving Innovation," Working Paper, Harvard University, MA.

Nordhaus, William D. (1991), "To Slow or Not to Slow: the Economics of the Greenhouse Effect," Economic Journal, 101 (407), pp. 920-937.

Nordhaus, William D. (1993), "Optimal Greenhouse Gas Reductions and Tax Policy in the 'DICE' Model," American Economic Review, May 1993 (Papers and Proceedings), 83 (2), pp. 313-17.

Nordhaus, William D. (1994), *Managing the Global Commons: The Economics of the Greenhouse Effect*, MIT Press: Cambridge, MA.

Nordhaus, William D. and Z. Yang (1996), "A Regional Dynamic General Equilibrium Model of Alternative Climate-Change Strategies," American Economic Review, September 1996, 86(4), pp.741-765.

Office of Technology Assessment (1991), "Changing by Degrees: Steps to Reduce Greenhouse Gases: Summary," OTA-0-482,, Washington D.C.: U.S. Government Printing Office.

Oliveira-Martins, J., J-M. Burniaux, and J.P. Martin (1992), "Trade and the Effectiveness of Unilateral CO_2-Abatement Policies: Evidence from GREEN," OECD Economic Studies, No. 19, pp.123-140.

Organisation for Economic Cooperation and Development (1997), *Evaluating Economic Instruments for Environmental Policy*, Paris: OECD.

Organisation for Economic Cooperation and Development (1996), *Climate Change, Economic Instruments and Income Distribution*. Paris: OECD.

Parry, Ian W.H. "Pollution Taxes and Revenue Recycling," Journal of Environmental Economics and Management, November 1995, 29(3), pp. 564-77.

Pezzey, J. (1992), "Analysis of Unilateral CO_2 Control in the European Community and OECD," The Energy Journal, 13, pp. 159-171.

Poterba, J.M. (1991), "Tax Policy to Combat Global Warming: On Designing a Carbon Tax." In Dornbusch, R. and J.M. Poterba, *Global Warming: Economic Policy Responses*. Cambridge, MA: MIT Press.

Reid, W. and J. Goldemberg (1997), "Slowing Climate Change: Is the South Doing More Than the North?" Draft Paper, World Resources Institute.

Repetto, R., R. Dower, R. Jenkins and J. Geoghegan (1992), *Green Fees: How a Tax Shift Can Work for the Environment and the Economy*, Washington D.C.: World Resources Institute.

Repetto, R. (1995), *Jobs, Competitiveness, and Environmental Regulation: What Are the Real Issues?*, Washington D.C.: World Resources Institute.

Richels, R., J. Edmonds, H. Gruenspecht and T. Wigley (1996), "The Berlin Mandate: The Design of Cost-Effective Mitigation Strategies," Energy Modeling Forum-14, Working Paper.

Richels, R. and P. Sturm, "The Costs of CO_2 Emissions Reductions: Some Insights from Global Analysis," Energy Policy, Vol. 24, Nos. 10/11, pp. 875-887.

Rosenthal, D., H. Gruenspecht and E. Moran (1995), "Effects of Global Warming on Energy Use for Space Heating and Cooling in the United States," The Energy Journal, Vol. 16, No. 2, pp. 77-96.

Rutherford, T. (1992), "The Welfare Effects of Fossil Carbon Restrictions: Results from a Recursively Dynamic Trade Model," OECD Dept. of Economic Statistics, Resource Allocation Division, Working Paper No. 112, OECD/GD (91), Paris.

Scheraga, J. and N. Leary (1993), "Costs and Side Benefits of Using Energy Taxes to Mitigate Global Climate Change," Proceedings 1993 National Tax Journal, pp. 133-138.

Shackleton, Robert, M. Shelby, A. Cristofaro, R. Brinner, J. Yanchar, L. Coulder, D. Jorgenson, P. Wilcoxen and P. Pauly (1992), "The Efficiency Value of Carbon Tax Revenues," U.S. Environmental Protection Agency.

Shelby, Michael, Robert Shackleton, Malcolm Shealy and Alexander Cristofaro (1995), "The Climate Change Implications of Eliminating US Energy and Related Subsidies," U.S. Environmental Protection Agency.

Terkla, David, "The Efficiency Value of Effluent Tax Revenues," Journal of Environmental Economics and Management, 11, pp. 107-123. (1994).

Titus, J (1992) "The Cost of Climate Change to the United States" in Majumdar, S.K., Kalkstein, L.S., Yarnal, B., Miller, E.W. and Rosenfeld, L.M. (eds.) *Global Climate Change: Implications, Challenges and Mitigation Measures*, Pennsylvania Academy of Sciences.

Tol, R.S.J., (1995) "The Damage Costs of Climate Change: Towards More Comprehensive Calculations," Environmental and Resource Economics, 5, pp. 353-374.

Triest, R. (1990) "The Effect of Income Taxation on Labor Supply in the United States," The Journal of Human Resources, Vol. 25, no. 3.

U.S. Department of Energy (1996), "Markal-Macro Model Documentation," Draft paper.

U.S. Energy Information Agency (1992), *Federal Energy Subsidies: Direct and Indirect Interventions in Energy Markets*. EIA/DoE Publication SR/EMEU/92-02, Department of Energy, Washington D.C.

U.S. Energy Information Agency (1996), *Annual Energy Outlook*, EIA/DOE 0383 (96), Department of Energy, Washington D.C.

U.S. Environmental Protection Agency (1996), "The Benefits and Costs of the Clean Air Act, 1970 to 1990," draft, Washington, D.C., May 3.

Wigley, T., R. Richels and J. Edmonds (1996), "Economic and Environmental Choices in the Stabilisation of Atmospheric CO_2 Concentrations," Nature, 379, pp. 240-243.

Worldwatch Institute, (1994) *Power Surge*, New York: W.W. Norton & Co.

Yin, E. (1993), Background Report for World Resources Institute, March 1993.

Yohe, G., J. Neumann, P. Marshall and H. Ameden (1996) "The Economic Cost of Greenhouse-Induced Sea-Level Rise for Developed Property in the United States," Climatic Change, Vol. 32, pp. 387-410.

Zollinger, P. and R. Dower (1996), "Private Financing for Global Environmental Initiatives — Can the Climate Convention's 'Joint Implementation' Pave the Way?," World Resources Institute, Washington D.C.

ANNEX

DETAILS OF THE STATISTICAL ANALYSIS

The data set is made up of 162 simulations derived from 16 models. From each simulation, the percentage change in CO_2 emissions and the percentage change in GDP, both relative to the baseline projections for the terminal year of the simulation, were recorded. In addition, a set of descriptive variables was recorded for each model simulation, representing basic structural assumptions and policy assumptions embedded in the model. These were represented either as binary dummy variables or as discrete numerical variables. All the variables considered in the analysis are presented in Table A.1 and include the following:

GDP	percentage change in real GDP relative to projected baseline, in terminal year
CO_2	percentage reduction in CO_2 emissions relative to projected baseline, in terminal year
MACRO	1 if a macro model, 0 if a CGE model
NCBACK	1 if there is a constant cost, non-carbon backstop
RECYCLING	1 if revenues from policy instrument are used to reduce existing distorting taxes
CLIMATE	1 if averted climate change damages are modeled
NON-CLIMATE	1 if averted air pollution damages are modeled
JI	1 if joint implementation or global emissions trading is modeled
PRODUCTION	1 if the model allows for product substitution
FUELS	the number of primary fuel types recognized for possible inter-fuel substitution
YEARS	the number of years available to meet the abatement target

Using the data, multiple regression analysis was performed to show how these modeling assumptions affected the predicted relationship between the change in GDP and the change in CO_2 emissions. The regression equation, with GDP as the dependent variable, was specified as a quadratic (through the origin) in the extent of CO_2 abatement from the projected baseline. The variables listed above were assumed to shift the coefficient of the linear term in CO_2, except that the variable for the non-carbon backstop was specified to affect the coefficient of the quadratic term in CO_2 because the presence of a backstop fuel source mainly affects the curvature of the abatement cost curve at high levels of abatement.

FORM OF REGRESSION:

$$GDP = \alpha_1 . CO_2 + \alpha_2 . (CO_2)^2 + \Sigma_{i=1}^{8} \beta_i . X_i . (CO_2) + \beta_9 . X_9 . (CO_2)^2$$

REGRESSION RESULTS

No. of observations: 162
R^2 : 0.83

	Coefficient	Standard Error
CO_2	-0.02319	0.00907
$(CO_2)^2$	-0.00079	0.00011
MACRO	-0.05548	0.01395
NCBACK	0.00051	0.00005
RECYCLING	0.04427	0.00652
CLIMATE	0.00943	0.00399
NON-CLIMATE	0.03823	0.00778
JI	0.02337	0.00327
PRODUCTION	0.00378	0.00365
FUELS	0.00018	0.00116
YEARS	0.00005	0.00006

Notes

1. Including an intercept term in the regression adds little to the regression's overall explanatory power. Moreover, the assumption that the regression passes through the origin is consistent with the models' definition of a baseline projection.
2. The inclusion of dummy variables to represent the individual models adds little to the explanatory power of the regression beyond that provided by variables representing underlying modeling assumptions.
3. The R^2 for a regression of GDP against CO_2 and $(CO_2)^2$ terms alone was 0.35.
4. Typically, the number of primary fuel types differs substantially from the number of energy sectors and energy technologies included in the models.

TABLE A.1 SUMMARY OF DATA

Model	Primary reference	Number of model runs	Years (range of end years)	Number of primary fuel types
BKV	Boyd, Krutilla and Viscusi (1995)	9	2000-2040	3
CRTM	Rutherford (1992)	9	2010-2100	7
DGEM	Jorgenson and Wilcoxen (1992)	24	2010-2050	3
DRI	DRI (1994)	12	2000-2020	7
Edmonds-Reilly-Barns	Edmonds and Reilly (1983, 1985)	14	2020-2095	7
EPPA	Yang et al. (1996)	8	2020-2100	5
Fossil2	AES (1990)	7	2000-2020	9
G-Cubed	McKibbin and Wilcoxen (1992)	9	2000-2020	2
Global 2100	Manne and Richels (1990a)	9	2010-2100	7
Goulder	Goulder (1995)	14	2010-2050	3
GREEN	Burniaux et al. (1990)	12	2010-2050	6
IIAM	Charles River (1997)	6	2010-2030	4
LINK	Kaufmann et al. (1992)	8	2000-2010	5
Markal-Macro	Hamilton et al. (1992)	3	2020-2030	11
MERGE2	Manne and Richels (1995)	12	2050-2100	7
SGM	Edmonds et al. (1993)	6	2010-2030	7

Notes

1. '1 or 0' signifies that different runs from the same model incorporate alternative assumptions.
2. The number of observations from each model varies with the number of alternative assumptions and end dates considered by the model.
3. Despite its name, the Markal-Macro model has optimizing characteristics and is not a macro model according to the distinction made in Box 1.
4. Because the Boyd et al. model is a comparative static model, the terminal years have been interpolated.
5. The results from the IIAM model come from the model's Single Open Economy Section of the model which does not include the effects of JI that appear in the Multi-Regional Trade (MRT) section.
6. The IIAM model is characterized as having no backstop because the backstop does not affect the time frame for which results were available.
7. Although the Edmonds-Reilly-Barns model has several non-carbon fuel sources, all are subject to increasing marginal costs, inconsistent with the definition of a backstop energy source adopted here (see Section 3.D for discussion).

SUMMARY OF DATA (CONTINUED)

Production substitution (1) or not (0)	Non-carbon backstop (1) or not (0)	Macro (1) or CGE (0)	Efficient revenue recycling (1) or lump-sum (0)	Joint implementation (1) or not (0)	Averted climate damages (1) or not (0)	Averted pollution damages (1) or not (0)
1	0	0	0	0	0	1
1	1	0	0	0	0	0
1	0	0	1 or 0	0	1 or 0	1 or 0
1	0	1	1 or 0	0	0	0
1	0	0	0	1 or 0	0	0
1	1	0	0	1 or 0	0	0
0	1	0	0	0	0	0
1	0	0	1 or 0	0	0	0
1	1	0	0	0	0	0
1	0	0	1 or 0	0	0	0
1	1	0	0	1 or 0	0	0
0	0	0	0	0	0	0
0	0	1	1 or 0	0	0	0
0	1	0	0	0	0	0
0	1	0	0	1	1 or 0	0
1	0	0	0	1 or 0	0	0

ADDITIONAL REFERENCES

The following references provided data and background information on models used in the analysis. This was in addition to several references in the main text.

Burniaux, J-M., J.P. Martin, G. Nicoletti and J.O. Martins (1991), "The Costs of Policies to Reduce Global Emissions of CO_2: Initial Simulation Results with GREEN," OECD Dept. of Economic Statistics, Resource Allocation Division, Working Paper No. 103, OECD/GD (91), Paris.

Burniaux, J-M., G. Nicoletti and J.O. Martins (1992), "GREEN: A Global Model for Quantifying the Costs of Policies to Curb CO_2 Emissions," OECD Economic Studies, 19 (Winter), pp. 49-92.

DRI/Charles River Associates (1994), "Economic Impacts of Carbon Taxes: Detailed Results," Research Project 3441-01 prepared for Electric Power Research Institute, November.

Edmonds, J.A. and J.M. Reilly (1985), *Global Energy: Assessing the Future*, Oxford University Press, New York.

Edmonds, J.A. and J.M. Reilly (1983), "Global Energy and CO_2 to the Year 2050," Energy Journal, 4 (3), pp. 21-47.

Edmonds, J.A. and D.W. Barns (1991) "Use of the Edmonds-Reilly Model to Model Energy-sector Impacts of Greenhouse Gas Emissions Control Strategies," prepared for the U.S. Dept. of Energy, by Pacific Northwest Laboratories, Washington, D.C.

Edmonds, J.A. and D.W. Barns (1990) "Estimating the Marginal Cost of Reducing Global Fossil Fuel CO_2 Emissions," Pacific Northwest Laboratories, PNL-SA-18361, Washington, D.C.

Edmonds, J.A., H. Pitcher, D. Barns, R. Baron and M. Wise (1993), "Modeling Future Greenhouse Gas Emissions: The Second Generation Model Description," Paper presented to the U.N. University Conference on 'Global Change and Modelling,' Tokyo, Japan, October.

Hamilton, Leonard D. et al. (1992), "Markal-Macro: An Overview," Brookhaven National Laboratory, Upton, N.Y.

Jorgenson, D. and P. Wilcoxen (1992), "Reducing U.S. Carbon Dioxide Emissions: The Cost of Different Goals," in *Advances in the Economics of Energy and Resources*, Vol. 7, pp. 125-158, Greenwich, CT.: JAI Press.

Jorgenson, D. and P. Wilcoxen (1990), "Intertemporal General Equilibrium Modeling of U.S. Environmental Regulation," Journal of Policy Modeling 12, pp. 715-744.

Kaufmann, Robert, H-Y Li, P. Pauly, L. Thompson, "Global Macroeconomic Effects of Carbon Taxes: A Feasibility Study", Paper prepared for the EPA, Washington D.C., Contract no. 68-W8-0113.

Kram T. (1994), "Boundaries of Future Carbon Dioxide Emission Reduction in Nine Industrial Countries," part of IEA Energy Technology Systems Analysis Programme Annex IV, Netherlands Energy Research Foundation.

Manne, A.S. (1992), "Global 2100: Alternative Scenarios for Reducing Carbon Emissions," OECD Dept. of Economic Statistics, Resource Allocation Division, Working Paper No. 111, OECD/GD (91), Paris.

Manne, A.S., and R.G. Richels (1995), "The Greenhouse Debate: Economic Efficiency, Burden Sharing and Hedging Strategies," The Energy Journal, Vol. 16, no. 4.

Manne, A.S. and R.G. Richels (1990b), "CO_2 Emissions Limits: An Economic Cost Analysis for the USA," The Energy Journal, Vol. 11, no. 2, pp. 51-74.

McKibbin, W. and P.J. Wilcoxen (1994), "The Global Costs of Policies to Reduce Greenhouse Gas Emissions," Final Report on U.S. Environmental Protection Agency Cooperative Agreement, CR818579-01-0, Washington, D.C.: The Brookings Institution.

Yang, Zili, Eckaus, R., Ellerman, A. and H. Jacoby (1996), "The MIT Emissions Prediction and Policy Analysis (EPPA) Model," Report No. 6, May, MIT Joint Program on the Science and Policy of Global Change.

ABOUT THE AUTHORS

Robert Repetto is Vice President and Senior Economist at the World Resources Institute and Director of the Program in Economics and Population. Formerly, he was an Associate Professor of Economics at the School of Public Health at Harvard University and a member of the economics faculty at Harvard's Center for Population Studies.

Duncan Austin is a Research Analyst in the Program in Economics and Population. Prior to joining WRI, he worked as an Economist in the Department of the Environment in the United Kingdom.

BOARD OF DIRECTORS

WORLD RESOURCES INSTITUTE

The World Resources Institute (WRI) is an independent center for policy research and technical assistance on global environmental and development issues. WRI's mission is to move human society to live in ways that protect Earth's environment and its capacity to provide for the needs and aspirations of current and future generations.

Because people are inspired by ideas, empowered by knowledge, and moved to change by greater understanding, the Institute provides—and helps other institutions provide—objective information and practical proposals for policy and institutional change that will foster environmentally sound, socially equitable development. WRI's particular concerns are with globally significant environmental problems and their interaction with economic development and social equity at all levels.

The Institute's current areas of work include economics, forests, biodiversity, climate change, energy, sustainable agriculture, resource and environmental information, trade, technology, national strategies for environmental and resource management, business liaison, and human health.

In all of its policy research and work with institutions, WRI tries to build bridges between ideas and action, meshing the insights of scientific research, economic and institutional analyses, and practical experience with the need for open and participatory decision-making.

WORLD RESOURCES INSTITUTE

1709 New York Avenue, N.W.
Washington, D.C. 20006, U.S.A.
http://www.wri.org/wri

WORLD RESOURCES INSTITUTE CLIMATE PROTECTION INITIATIVE

The Climate Protection Initiative (CPI) is a partnership between WRI and private firms to identify acceptable policies and business strategies for achieving strong climate protection goals. This institute-wide effort is motivated by the belief that there is a positive link between climate protection and a sound economy. Wider recognition of this linkage is essential for the development and implementation of an effective international climate treaty.

WRI's CPI team is working with leading corporations to define Safe Climate, Sound Business Scenarios——policy pathways for stabilizing greenhouse gas concentrations and meeting world energy demand, while maintaining a healthy economic environment. We are contributing to business strategies by developing case studies of how companies have overcome barriers to implementing carbon dioxide reduction initiatives and achieved financial and productivity benefits. And, we are investigating how the development of advanced technologies could position such industries as electronics and communications as leaders in climate protection.

The build-up of greenhouse gases in the atmosphere and human-caused warming of the Earth is the largest global environmental challenge facing society as it approaches the 21st century. According to the Intergovernmental Panel on Climate Change, if action is not taken to slow and reverse the growth of greenhouse gas emissions, there could be ecological impacts on a scale outside all of human experience. In turn, these impacts would have serious economic and social consequences. Significant technological, economic, and policy changes are needed now and in the coming decades to head off these risks.

The actions needed to address the threat of climate change, however, provide an opportunity to chart a more sustainable course into the future as individual firms and national economies become more efficient, embrace environmentally-sound technologies, and accelerate the transition to a global economy fueled by renewable energy resources.

The CPI team is committed to designing and advancing policy options that are flexible and market-based, allow least-cost mitigation strategies, and are sensitive to competitiveness issues within and among nations. Through this initiative, WRI seeks to reduce the risks of climate change while nurturing sustainable economic development worldwide.

For additional information on the Climate Protection Initiative, visit WRI's website at:

http://www.wri.org/wri/climate/

or write to:

Climate Protection Initiative
World Resources Institute
1709 New York Avenue, N.W.
Washington, D.C. 20006, USA